THE EXPLORATION OF MARS

The Mars expedition, 8600 miles from its goal. The ships are approaching tail first.

THE EXPLORATION OF
MARS

by WILLY LEY and WERNHER VON BRAUN
with 16 paintings in color and 5 in black and white

by CHESLEY BONESTELL

COMMONWEALTH BOOK COMPANY
ST. MARTIN, OHIO

ISBN: 978-1-948986-61-8

CONTENTS

LIST OF FIGURES

LIST OF PLATES

LIST OF PLATES

THE EXPLORATION OF MARS

1. RED PLANET IN THE SKY

IT WAS during the first year of our century. The northern hemisphere of the planet earth was experiencing winter, the winter bridging the years 1900 and 1901. When the night skies were clear a bright reddish star could be seen in the midst of the bluish-white lights of the familiar constellations. That bright star was not always there, but the people who saw it shining down on the snowy winter landscape knew well what it was, for their newspapers and magazines had told them about it in advance. This was the planet Mars, once more near earth and still coming closer. The papers had told the date too; it would be closest to earth late in February of 1901. Afterward it would slowly grow fainter and fainter, finally becoming invisible to the untrained and unaided eye. But it would be back again in the sky in the spring of 1903.

Teachers pointed out the red planet to their pupils; fathers took their sons out into the open or up on the roof of the house late at night to look at it. And almost for the first time in history people looked up to the red planet without fear. Instead of being faintly uneasy at the thought of having Mars, the symbol of the god of war, hanging in the sky, they regarded it with intense curiosity, mixed with a strange kind of altruistic hopefulness. Instead of looking forward to news of strife and ominous unrest, people waited for news from a source which had rarely been considered newsworthy in the normal course of events: the astronomical observatories. The men who did not just look at the red planet but observed it with the best and most powerful instruments at their disposal at the time would soon have news to report. And virtually everybody was convinced in advance that the news would be interesting and, in a highly idealistic sense, "good."

The attitude of looking up into the sky at Mars with a longing hopefulness instead of with fear was quite recent, and it was the unequivocal result of scientific work, plainly and directly traceable to statements made by men of science,

3

especially astronomers, during the preceding two decades. Up until then fear and uncertainty had been the result of having Mars in the sky. Its rather sudden, though predictable, appearance at intervals meant to earlier human beings that they were not, and could not be, master of their fate, that they were subjected to inescapable influences. And the red color reminded them of fire and blood. Just as soon as humanity had learned to write, the name of the red light which flamed brightly and ominously for a few months every second year was written down. To the Egyptians it was Harmakhis. To the Chaldean stargazers it was Nergal—and Nergal was also the name of the god of battle and of the dead.

As human history progressed and other peoples, speaking other tongues, assumed importance, the names for the red-colored celestial visitor changed, but the idea first introduced by the Chaldeans did not. In Persia the planet was Pahlavani Siphir, the Celestial Warrior. To the Greeks it was Ares. Etymologists are not certain whether Ares is derived from ἄρω (to kill) or from ἀρά, which translates as disaster or even vengeance, but the general meaning cannot be in doubt. And Roman Mars, of course, was the god of war too. Even the symbol for Mars (♂), which is used by astrologers to impress clients, by astronomers when they are crowded for space, and by biologists, especially botanists, if they want to indicate that something is of male sex, is believed to be a military symbol, composed of shield and spear.

Like the other major gods of antiquity Mars was given a day of the week in his honor: *Martis dies.* The name survives in virtually unchanged form in French as *Mardi,* and our word Tuesday carries the same idea, with the Germanic god of war substituted for the Latin Mars. Since the god of war was not only to be respected but also to be watched closely, it is not surprising that the planet Mars is mentioned again and again in ancient writings. According to the French astronomer Eugenious Michael Antoniadi, a whole cuneiform treatise on the motions of Mars has been found at Nineveh. One could watch, but one could also try to placate and appease, an attitude which found perhaps its strongest expression in the originally Persian and Chaldean religious sect of the Sabians. They met on Tuesdays, dressed in red, smeared with blood, and carrying naked swords or daggers. The sect, and its ideas, spread as far westward as Rome. Finally Christianity wiped out the star worship, but, as the Romans would have said, *semper aliquid haeret* (something always remains), for neither the appearance of Mars in the sky nor its supposed manifestations on the ground ever ceased. And anybody sufficiently superstitious could point out as late as 1900 that Mars had been close in the war years of 1811, 1813, 1864, 1871, and 1899, and that the Boer War, which began in the last-mentioned year, was still going on.

But to the vast majority of the people who looked for Mars in the wintry night sky of February 1901 all this was ancient history, interesting to know, no doubt, but of historical interest only. Certainly, humanity had its full share of strife and struggles, but up there, on that other planet, moving around the sun just like the earth but needing about two terrestrial years for one complete revolution because it was farther away from the sun—on that other and smaller planet there lived another, probably older, and in any case wiser, humanity which had evolved beyond internecine strife. Astronomers had sent from their observatories to the popular press reports and drawings which seemed to show colossal works of engineering, still beyond human capacity to build and certainly beyond human ability for co-operation. Incidentally, such feats of engineering were not, or not yet, needed on earth, but there seemed to be proof that, *if* needed, they could be accomplished.

Some two decades earlier word had come from Italy that the director of the observatory of Milan, Giovanni Virginio Schiaparelli, had seen a network of fine lines on the surface of Mars. They appeared, the report ran, to be absolutely straight and quite long, many hundreds of miles long. Schiaparelli had used the opportunity of a close approach of Mars and the clear seeing afforded by the fine sky over Milan to map these lines. He had called the lines *"canali"*; the Italian word means "channels" or "grooves," but it seemed obvious to call them "canals."

To a well-read and open-minded man at the turn of the century this discovery, first reported from Italy and then confirmed by observers in the United States as well as in France, was really just a final proof for a philosophy that had grown strong on the discoveries of nineteenth-century science. It was a philosophy stressing something now taken for granted—namely, the Unity of Nature—as exemplified by a long chain of proofs that had begun with Friedrich Wöhler's *synthesis of urea* in a test tube. By this chemical feat Wöhler had shown that there was no fundamental difference between so-called "organic" and so-called "inorganic" chemical compounds; the same elements composed both; the same atoms could go into either the make-up of a flower or the black soil from which it grew. The next major link in that chain had been Kirchhoff's and Bunsen's development of spectrum analysis, which proved that other celestial bodies, too, consisted of the same elements we knew from our terrestrial environment and which compose our own bodies. Charles

5

Darwin's theory of evolution was another link in the same chain, showing that differences among living things were also differences of degree rather than fundamental ones, leading through uncountable numbers of forms from the most primitive to the most advanced. And since the most primitive form of life must ultimately have sprung from compounds which were not normally considered alive, it had to be concluded that life could not be restricted to earth but that it had sprung up—and was still springing up—wherever conditions in the universe were favorable.

It was of course a particularly lucky accident, but essentially in the nature of a demonstration, that the planet that is almost our nearest neighbor in space showed visible proof of life, and not just of life in general but of intelligent life.

One could read the news about Mars in many books and in many languages. In France an astronomer named Camille Flammarion, author of many works, had published his enormous *La Planète Mars et Ses Conditions d'Habitabilité* in 1892, and six years later he had drawn a globe of the planet, showing its reddish-yellow deserts, its bluish-green dark areas, and its canals, which could be bought in most stores where ordinary globes were sold. In America Percival Lowell had published in 1895 a book entitled simply *Mars,* in which one could find a step-by-step development of the concept of an intelligent race on Mars, politically unified and distributing the rare and therefore valuable water supply over planet-wide distances. In Germany Professor Jakob Heinrich Schmick of Cologne had published a book called *Der Planet Mars, eine zweite Erde* (*The Planet Mars, a Second Earth*) in 1879, only two years after Schiaparelli had seen those canali for the first time. As a scientist Professor Schmick specialized in the climatology of the earth's past; as a philosopher he held an opinion which was perfectly expressed in the title of his book. In Austria a man named Otto Dross published (in 1901) a book entitled *Mars, eine Welt im Kampf ums Dasein,* which was a paraphrasing of a term coined by Charles Darwin, reading, in translation: *Mars, a World Engaged in the Struggle for Survival.* In 1892 several observers had reported that a new dark area had formed on Mars; Dross pointed to this as proof that the Martian engineers had succeeded in wresting another piece of land from the desert and making it bear fruit.

And there were not very many people at that time who were not aware of the Prix Pierre Guzman, which had been announced by the French Academy of Sciences on December 17, 1900. In dry and somewhat legal language the Academy announced that Madame Clara Goguet, the widow Guzman, had deposited with the

Academy the sum of 100,000 francs for an award to be known in memory of her son as the Pierre Guzman Prize, to be given to whoever succeeded in establishing communication with another world *other than Mars!* [1] Apparently the widow Guzman thought that communication with Mars would be too easy to deserve a prize. Camille Flammarion, whose book had inspired the whole, withdrew with some horror. This, he said, this exclusion of Mars, is *une idée bizarre,* since it renders *hors de concours* the only planet which "seems to be in a situation to participate."

The only saving grace, as far as astronomers were concerned, was that Madame Goguet foresaw that the prize "might not be awarded quickly" and had decreed that the interest on the capital of 100,000 francs should be paid out every five years, as a secondary prize, for any important contribution to astronomy and that this secondary prize should be paid for the first time in 1905. (It was awarded to the director of the observatory at Nice, Henri Joseph Anastase Perrotin, and paid to his widow because Perrotin died in 1904.)

Possibly with his eye on the Prix Guzman, either the main or the secondary award, Monsieur A. Mercier, in 1902, organized a symposium on the planet Mars which was held in the Hotel de Ville of the city of Orléans. The founding of a society for communication with the planets was advocated at that symposium, but the proposal did not get anywhere. The experts shook their heads. Much as they agreed with the trend of thought, no feasible method of establishing communication was known. Joseph Johann von Littrow of the Vienna Observatory had once advocated the use of geometrically arranged fires as signals for the inhabitants of other planets. Aside from the expense involved, it seemed impossible to produce signals of sufficient size. The inhabitants of the planet Mars, several savants mused, would have a much easier task if they wanted to do the same. Since we see the daylight side of Mars when it is closest, all they would have to do was to form large geometrical symbols of mirrors or shiny sheet metal and lay them on the ground; the sun would then produce strong optical signals. And though those "electrical waves" discovered and demonstrated by Heinrich Hertz might one day find some application in the field of communications, not enough was yet known about them to foretell what they might or might not be able to do. Communication with Mars had to be postponed, pending later inventions and discoveries.

1. *Mme Clara Goguet, veuve Guzman, a légué, à l'Académie des Sciences, une somme de cent mille francs pour la fondation d'un prix qui portera le nom de Prix Pierre Guzman, en souvenir de son fils, et sera décerné à celui qui aura trouvé le moyen de communiquer avec un astre autre que la planète Mars. . . .*

THE EXPLORATION OF MARS

All one could do in the meantime was to observe, watch for changes, and confirm or clarify beliefs. One day, perhaps . . .

It had taken thousands of years for humanity to progress from the idea of Nergal—a noncorporeal and dimensionless light in the sky—to the concept of the "sister planet and neighbor in space." Most of the changes in knowledge and attitude had been compressed into three centuries but the beginnings had been much farther back in time.

Of the observations of Mars in classical times only one is worth mentioning, not only because it is a rather rare case but also because of the stature of the man who made it and recorded it, who happens to be Aristotle. During Aristotle's lifetime it happened that the moon and Mars, as seen from the earth, formed a straight line so that Mars disappeared behind the moon. Such an event is known in astronomical language as an occultation. Aristotle drew the correct conclusion from the one he observed: since Mars disappeared behind the moon it was "higher up in the heavens," or farther away. By drawing this conclusion Aristotle made the first tiny step in the intellectual exploration of Mars. At least it was now known that Mars was more distant than the moon.

The second step, namely, the recognition that Mars was a world in its own right rather than just "a luminosity," took a long time. Long before it happened Aristarchus of Samos made the grandiose—and correct—guess that the stars which the Greeks called *planetes*, or "wanderers," moved around the sun instead of around the earth, and that the earth was in fact one of these *planetes*. Another Greek astronomer did not agree with this thought, however. He was the great Hipparchus, who catalogued more than a thousand stars and who invented the concept of longitude and latitude to fix the position of points on earth. He put the earth back into the center of the universe. More than two centuries later Claudius Ptolemaeus confirmed this opinion and was credited by later generations with having invented it, so that they spoke of the Ptolemaic system. But whether a philosopher bowed to Ptolemaeus or considered that Aristarchus might have some arguments in his favor, Mars remained a mere light in the sky.

To give it body required progress in an entirely different field of human endeavor. The decisive step was taken in the Netherlands during the earliest years of the seventeenth century. The spectacles-maker Jan Lippershey invented the telescope, and news of his achievement spread over Europe. In Italy Galileo Galilei, among other things a member of the *Accademia dei Lincei* (the "Academy of the Lynxes," to our notions a strange name for a society for the advancement of science), heard that the invention "made distant things appear near" and that it involved several spectacles lenses. After a little reasoning and a very short period of trial and error Galilei built his first telescope (he still called it *perspicillum* when writing in Latin and *occhiale* when writing in Italian) and looked through it at the sky. A host of discoveries fell into his lap and delighted the other "lynxes," but the fundamental one was this: even with the new instrument the fixed stars remained pinpoints of light, but all the planets showed disks. One might still argue about the stars; the planets were bodies.

In the years immediately preceding the invention of the telescope and the first harvest of astronomical discoveries, the story of Mars, and of the planets in general, had described an interesting curlicue. That Aristarchus had wanted to place the sun in the center and to have the planets, including the earth, move around it was still—or again—known to educated people with an interest in astronomical matters. One of those who knew about this theory was the Danish nobleman Tycho Brahe. He did not believe it, and this very disbelief caused him to find himself in a dilemma, because the other system, still called the Ptolemaic system, did not satisfy him either. In the Ptolemaic system the earth was in the center, and all the planets, among which both the sun and the moon were counted, moved around it. It was taken for granted that the heavenly bodies moved along a curve that was considered "philosophically perfect"—namely, the circle. Unfortunately the observations did not quite agree with this simple picture.

To make movement along circles agree with what could be seen and measured, all kinds of corrections became necessary. In the first place, the center of all these circles could not be the center of the earth. Since the earth was the obvious and undisputed center of the universe, the center of the circles along which the planets moved was *not* the center of the universe. Besides, the planets could not move along their circles directly. They moved along a small circle the center of which moved along the big circle. Of course the size of the smaller circle—technically known as epicycle—differed for each planet. Even so, things did not quite work out; one day one might even have to accept the notion of irregular movement. Tycho Brahe did not like any of this. He would not accept Aristarchus, because the earth on which he stood was solid and large and heavy. The moon might be solid too, but at least it was much smaller. To Tycho Brahe—remember, this was before the invention of the telescope—the planets were just balls of light without weight.

It was probably this idea of "weight" which suggested what later came to be called Tycho's strange compromise. The planets, he decreed, moved around the sun. But the sun, accompanied by the planets, moved around the heavy earth. After the idea had been conceived it had to be proved. For that Tycho Brahe needed observations by the thousand, and they had to be observations with precise measurements. One could make such measurements in pretelescopic times by sighting on a star, or a planet, along a straight stick, or along a string, and measuring the angle formed by the string with the ground. If an observation was not precise enough, Tycho reasoned correctly, it was because the instrument was too small. He built larger ones, and still larger ones, and he did amass the thousands of observations he wanted.

The book on the *Revolutions of the Celestial Bodies* by Nicholas Copernicus of Thorn had already been published. Copernicus had stated, with specific reference to Aristarchus, that the sun was in the center and that the planets moved around it, but he had not changed the classical concepts of the circular paths with epicycles on them. One might say that he had merely made sun and earth change places in the Ptolemaic system. Tycho Brahe obviously did not like any of this. Near the end of his life he had the heaps of careful observations which he had always wanted and knew well that they were the raw material for a proof of how the solar system was constructed.

For a while Tycho Brahe had an assistant, a young mathematician with interesting though somewhat mystical ideas. His name was Johannes Kepler. Both Tycho Brahe and Kepler were supported financially—though not very well—by Rudolph of Habsburg, then Holy Roman Emperor. When Tycho died, Kepler not only succeeded to his official position, he also fell heir to all the observations and to the job of analyzing them. One thing that did not trouble him was "Tycho's compromise." Kepler had been convinced all along of the essential correctness of Copernicus's ideas and of the position of the sun as the center of the system. He went even further and assumed that the force which kept the planets moving originated in the sun. If this were so, the planets nearer the sun should move faster than the planets farther away. It seems—mostly because Kepler made this point very strongly—that others had said or implied that the different periods of revolution were due to the length of the orbit to be traveled. It was easy for Kepler to see that the planet closest to the sun, Mercury, actually moved along its orbit faster than did the next planet, namely Venus. And Venus moved faster than earth, which travels an orbit enclosing that of Venus, and earth, in turn, was faster than Mars, which has an orbit enclosing that of earth.

The question was what shape these orbits could possibly have. Since, in Kepler's opinion, the moving force originated in the sun, it had to act on the planets directly; he could not have epicycles. And without epicycles, as astronomers had known for a thousand years and more, a circular orbit simply did not agree with observation. The way to solve the problem was to investigate a particular orbit down to the minutest detail. Kepler picked Mars as his example. For some time he thought that the path of Mars around the sun might be oval in shape, and he spent years trying to work out a reasonably easy method of dealing mathematically with an oval. Finally he gave up and looked for another closed curve. An ellipse might be the answer.

To anybody but a mathematician there seems to be a rather pronounced difference between a circle and an ellipse. But to a mathematician the circle is merely a special case of an ellipse. In a circle every point of the periphery is at the same distance from the center. In an ellipse every point of the periphery is related to two focal points, and the sum of the distances to both points is always the same. If these two focal points are far apart from each other, the result is a rather elongated and narrow ellipse. The closer the two focal points come to each other the "rounder"— technically, less eccentric—the resulting ellipse. And if the two focal points fall together a circle is the result.

Kepler found that the orbit of Mars is an ellipse and that the sun occupies one of the two focal points of this ellipse. The other focal point is empty. This simple statement is what is now called Kepler's First Law. During the mathematical work on the orbit of Mars—Kepler stated that he made the calculations *seventy times* to be certain—it had become quite obvious to him that Mars did not move with a uniform velocity along its orbit. This fact strengthened his belief that the moving force was in the sun, for Mars moved more slowly when in that portion of its elliptical orbit which was farthest from the sun and noticeably faster when in the portion closer to the sun. Kepler even succeeded in finding a way of expressing the changing velocity of the planet in a simple manner. He drew a line from the sun to the planet and called this line, which more or less corresponded to the radius of a circle, the "radius vector." When the planet was near the sun it moved through a larger angle in a given time, say, in 24 hours, and when far from the sun it moved through a lesser angle. But when the planet was far from the sun the radius vector was longer, and in any case, whether near or far, the areas covered by the radius vector in a given time were equal. This became Kepler's Second Law: *"The radius vector sweeps over equal areas in equal times."* It was now known *how* the planet Mars moved, and of course what applies to Mars applies to all other planets too.

Kepler published his findings in 1609, one year before Galilei published his *Sidereus nuncius (Messenger of the Stars)* which reported on the first telescopic observations. The main title of Kepler's book was *Astronomia nova (New Astronomy)*, but the equally important subtitle was *De Motibus Stellae Martis (On the Motions of Mars)*. The official place of publication was Prague, where Kepler lived and where Rudolph of Habsburg spent most of his time, but the book was printed in Heidelberg. The Emperor considered the book his personal property, since he had paid Kepler's salary at the time, or at least had promised to do so. He forbade Kepler to send copies of the book to other astronomers and mathematicians without his knowledge and permission,[2] and circulation of the work was very poor at first. Kepler would probably have been content to bow to the Imperial will if he had received his salary. Since he did not, he finally sold the whole edition to the printer, who then sold single copies to recoup his money.

The book is very difficult to read, and it seems that Galilei, who did receive a copy, never took the time to read it. Others found the going hard too. The astronomer P. Crüger of Danzig, now mostly known as the teacher of the more famous Johannes Hevelius, wrote to his friend Professor Philippus Müller in Leipzig that "the Mars book requires a reader who is not troubled by any other thought, and not just for a day but for a full year." Apparently he succeeded in freeing himself from other thoughts enough to read his way through slowly, for several years later he wrote, "I now no longer reject the elliptical shape of the planetary orbits and have convinced myself by the proof contained in Kepler's Mars book."

Since both the earth and Mars travel around the sun in orbits of different lengths and with different velocities their distances from each other vary constantly. And every 780 days (be it noted at once that this is an averaged figure and that the actual time interval may be as short as 765 days or as long as 810 days) the planet earth overtakes its slower neighbor Mars in the unending race around

2. "*. . . dass er* [Kepler] *one Unser vorwissen und bewilligung nymanden kain Exemplar davon gebe*" is the wording of the original document.

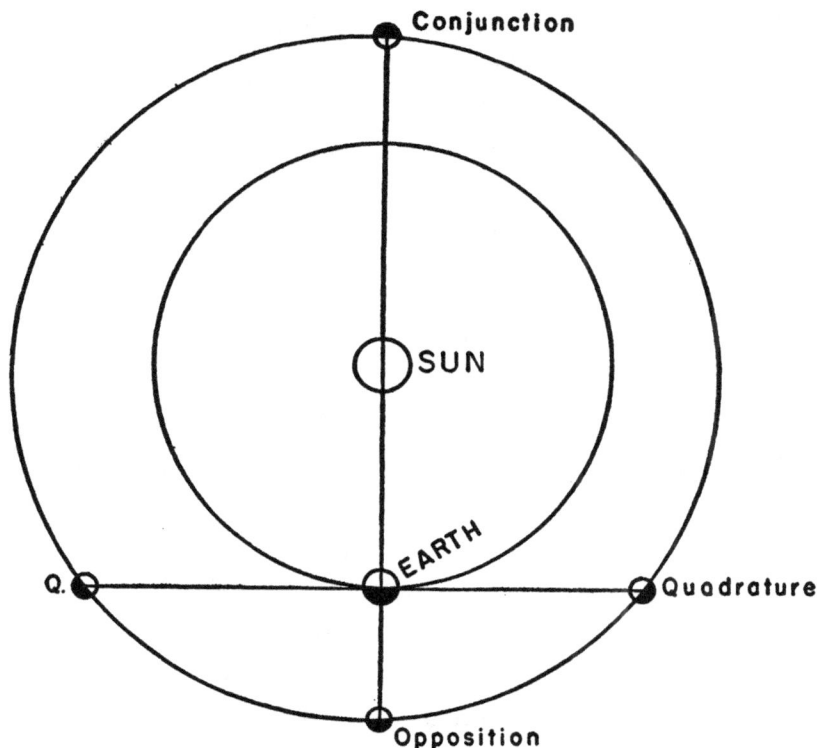

FIG. 1. The four principal positions that Mars can assume relative to earth. At conjunction it becomes invisible in the rays of the sun. The main period for observations is about from quadrature to quadrature.

the sun. Seen from the sun, the two planets will then momentarily lie along a straight line originating in the sun's center. Seen from the earth, the planet Mars is then in a spot directly opposite to that occupied by the sun in the sky—of course it is impossible to see both of them simultaneously because the bulk of the earth is in the way. The astronomers of the past used to say, when that happened, that Mars was in opposition—to the sun, that is. The term is still in use, though the connotations have been lost. We are not too much interested in the position of Mars in the sky relative to that of the sun; what we are interested in is the fact that Mars is then especially near to earth.

Because the earth keeps racing ahead and Mars is steadily falling behind, there are also times when the two planets form a straight line with the sun in the middle. Seen from the sun the two planets are then in opposition to each other; seen from the earth the planet Mars disappears from view in the rays of the sun and is said to be in conjunction. To us the importance of this is that Mars is as far away as it

13

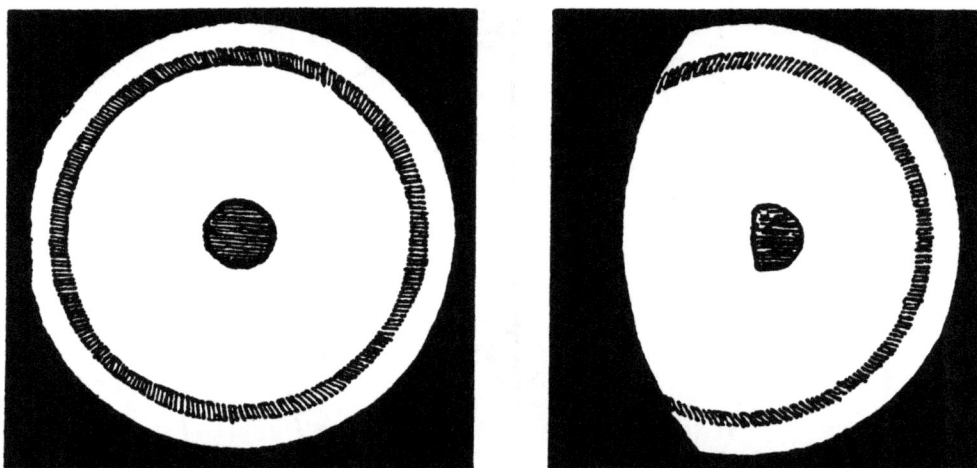

FIG. 2. The first telescopic drawings of Mars, made by Francesco Fontana in 1636 (*left*) and in 1638 (*right*)

can possibly be: from the earth to the sun it is 93 million miles, plus another 93 million miles from the sun to the opposite point of the earth's orbit, plus another 50 million miles from that point to the orbit of Mars.

In addition to conjunction, when Mars, sun, and earth form a straight line with the sun in the middle, and opposition, when they also form a straight line but with the earth in the middle, there is one other position that is of some importance. If, as seen from the earth, the sun and Mars are at right angles to each other, they are in the position called quadrature. This is the only time that the planet Mars does not look perfectly or very nearly round to us. As a rule, since it is farther away from the sun than we are, we see a fully illuminated disk; we always look at the Martian hemisphere which happens to be in daylight. But when near quadrature we see a portion of its night side—or rather we don't see it, since the blackness of the night side of Mars blends in with the black sky. Mars then appears in the telescope as a gibbous moon.

The first man to make a telescopic drawing of Mars caught this gibbous phase on one occasion (Fig. 2). He was Francesco Fontana of Naples. His first picture was drawn in 1636 when Mars was in opposition. Fontana neglected to state the day on which the drawing was made, or else forgot what the date had been between the time the drawing was made and the time his book was published, which was in 1655. The second drawing, he noted, had been made on August 24, 1638. Mars was in quadrature then, but Fontana drew the phase presented by the planet about

14

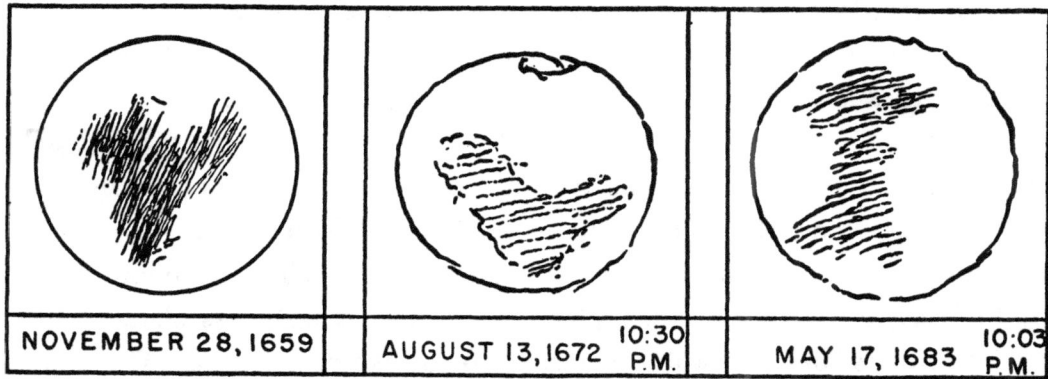

FIG. 3. Three drawings of Mars by Christiaan Huyghens

twice as pronounced as it actually is. Unfortunately, Fontana's telescope must have been a very poor instrument, for the Martian features which appear in his drawings —the darkish circle and the dark central spot which he called "a very black pill"— obviously originated inside his telescope. His drawing of Venus also displayed a "black pill" in the center.

The earliest telescopic drawing of Mars to show a definite surface feature is the one made by Christiaan Huyghens in 1659. Huyghens, son of a Dutch diplomat and poet, was an accomplished experimental physicist and mathematician. His astronomical fame is partly based on the fact that he, in collaboration with his brother, invented a new method of grinding lenses. He used these lenses in his own telescopes, and they were good enough to show clearly something not known at the time—that Saturn's rings do not touch the body of the planet at any point. In the process of checking on the rings Huyghens discovered Saturn's largest moon, Titan. And in the course of time he produced several good drawings of Mars. The first one shows what is now called Syrtis major, a dark triangular marking with its point toward the Martian north pole, which, in all astronomical drawings and photographs, is located at the bottom of the disk. Huyghens' drawing of 1672 is considered the earliest to show the southern polar cap of Mars clearly.

The two books usually mentioned in connection with Huyghens' name are his work on clocks, entitled *Horologium oscillatorium*, and his optical work, the *Treatise on Light* (*Traité de la Lumière*), but his posthumous *Cosmotheoros*, which appeared in 1798, should not be forgotten. It is mentioned occasionally in astronomical works, but the stock remarks used give the impression that the authors just paraphrased earlier books using similar stock remarks, without going back to

15

the book itself. But the *Cosmotheoros* is still worth reading; it is a highly logical, charming, and often witty speculation about the inhabitants of other planets.

And it is surprising how many "modern" ideas one can find in the *Cosmotheoros*.[3] Huyghens did not waste any time speculating about the inhabitants of the moon, for the moon "has no seas, no rivers, nor clouds, nor air and water." Huyghens also knew that the moons of Jupiter and Saturn, like our own moon, always turn the same hemisphere toward their planets. The most important question, to Huyghens, was whether a planet had water; a waterless planet could not have inhabitants "because water is needed to partake of nutrition." Then he drew up a list of what he considered necessary for the existence of inhabitants, and also a list of the traits he considered as defining "inhabitants." There must be air "to carry sound to their ears." They must have ears and eyes—in short, senses, and they must have pleasure brought to them by their senses. As regards their size, it is not indicated by the size of the planet. All this could as well apply to mere animal life; the term "inhabitants" implied a difference similar to that between men and animals on earth. "Men differ from beasts in the study of Nature," Huyghens declared, and went on to list particulars for "inhabitants":

"They must have geometry and arithmetic and writing."

"They must have hands to make things."

"They must have feet to move around."

"They must have houses for protection from weather."

He also thought that "they must be upright," to free their hands for the making of things, but "it follows not therefore that they have the same shape with us," because a rational soul may inhabit another shape.

Being obviously preoccupied with Mars, Christiaan Huyghens had no love for astrological symbolism. At one point he made the caustic remark, "Saturn and Mars, I don't know why, but all fortunetellers hate them"—in horoscopes these were always the two "evil planets." For the same reason he made some equally caustic if most pertinent remarks about Father Athanasius Kircher's book about the solar system, published in 1656 under the title *Iter ecstaticum coeleste*, which may be translated as *Ecstatic Voyage*. As for Father Athanasius Kircher himself, he was a prolific writer on many subjects and had been professor of mathematics and Hebrew at the College of Rome, a position he gave up to devote his full time to the

3. An English edition appeared in Glasgow very soon after the publication of the original. Its title page read: *Cosmotheoros: or, Conjectures concerning* *the Inhabitants of the Planets.* Translated from the Latin of Christian Huygens.

study of hieroglyphics and archaeology. He is also credited with having invented the magic lantern. But when he started writing about astronomy he was handicapped by the facts that he was a Catholic scholar and a Jesuit. Because of involved politics within the Church, Galileo Galilei had been forced to recant his teachings that the earth moved, the fundamental volume of Copernicus had been put on the Index of forbidden books, and Kircher had to defend a position which would still have been defensible in 1606 but no longer was in 1656. He could not endow the earth with any motion, not even diurnal rotation, and he had to adhere to the position that creation had been confined to earth, so that there could be neither plants nor animals, and especially no "inhabitants," on the other planets. All he could describe were lifeless landscapes of religious or astrological significance. But he could accept the notion of some Arabs that the fixed stars were suns with planets. Huyghens pointed out with sarcastic glee that all these suns, with their planets, had to whirl around the earth once every 24 hours, and remarked that Kircher, if the Inquisition did not permit him to write what he thought, would have done much better not to have written the book at all.

Mars, in Kircher's book, appeared as its astrological symbol in a description which Huyghens summed up with the words "nothing but infernal, stinking, black flames." Against this he set his own description:

Mars . . . has some parts of him darker than other some. By the constant returns of which his nights and days have been found to be of about the same length with ours. But the inhabitants have no perceivable difference between summer and winter, the axis of that planet having very little or no inclination to his orbit [4] as has been discovered by the motion of his spots. Our earth must appear to them almost as Venus does to us, and by the help of a telescope will be found to have its wane, increase, and full like the moon; and never to remove from the sun above 48 degrees, by whose discovery they see it, as well as Mercury and Venus, sometimes pass over the sun's disk. They as seldom see Venus as we do Mercury. I am apt to believe that the land in Mars is of a blacker colour than that of Jupiter or the moon, which is the reason of his appearing of a copper colour, and his reflecting a weaker light than is proportionable to his distance from the sun. His body, as I observed before, though farther from the sun, is less than Venus. Nor has he any moons to wait upon him, and in that, as well as Mercury and Venus, he must be acknowledged inferior to the earth. His light and heat is twice and sometimes three times less than ours, to which I suppose the constitution of his inhabitants is answerable.

During Huyghens' lifetime—and partly by Huyghens himself—another important forward step was made: the period of the axial rotation of Mars was

4. Actually the axis of Mars is tilted to a slightly higher degree than that of earth (see Chapter 2).

determined. Huyghens entered in his diary under the date of December 1, 1659: "The rotation of Mars seems to take 24 terrestrial hours like that of earth." This first approximation was corrected in 1666, when Mars was in opposition, by another great name in the history of astronomy. This was Giovanni Domenico Cassini, or, as he called himself after becoming a French subject, Jean Dominique Cassini. He was Huyghens' contemporary; in fact the life spans of the two men coincided closely, except that Cassini lived longer. Cassini went to France at the invitation of the king to become the first director of the Paris Observatory, and might be said to have founded a dynasty of French astronomers: from 1700 to 1800 there was always a Cassini in an important astronomical position; only one near the end of the line deserted the fold to become a botanist.

During the opposition of Mars of 1666 the first Cassini was still professor of astronomy at Bologna, and, with the aid of observers in Rome, he determined the period of rotation of Mars (see Plates I and II). By timing carefully the positions of the markings of Mars he found that Mars presented the same picture 40 minutes later on each successive night.

In 1672, during another opposition, Giacomo Filippo Maraldi, a nephew of Cassini, observed Mars specifically for the purpose of confirming, or else of correcting, this result. He did not make a correction then, but continued to observe, and when the opposition of 1704 occurred he revised his uncle's figure downward by 1 minute. (Two and a half minutes would have been nearer the truth.) By 1704 Maraldi was also sure that the white areas marking the poles and the darker areas of the equatorial regions changed from opposition to opposition. After the opposition of 1719 Maraldi announced that he had actually observed changes to take place during the period of opposition. He cautiously said that what he observed might have been changes in cloud formations. In addition to this discovery, Maraldi reported that the white spot on the Martian south pole, though round, was not centered on the geographical pole. With a caution that possibly was carried a bit too far, Maraldi did not speak of an ice cap, or even of a snow cap, but referred simply to the "*tache blanche*" (white spot), without making a guess as to its nature. But he did say that it was eccentric to the pole—as are the earth's polar caps, a fact that probably was still unknown in 1719.

Half a century went by during which nothing was added to human knowledge of Mars. But then came a set of favorable oppositions and a new observer with new instruments. The oppositions were those of 1777, 1779, 1781, and 1783. The ob-

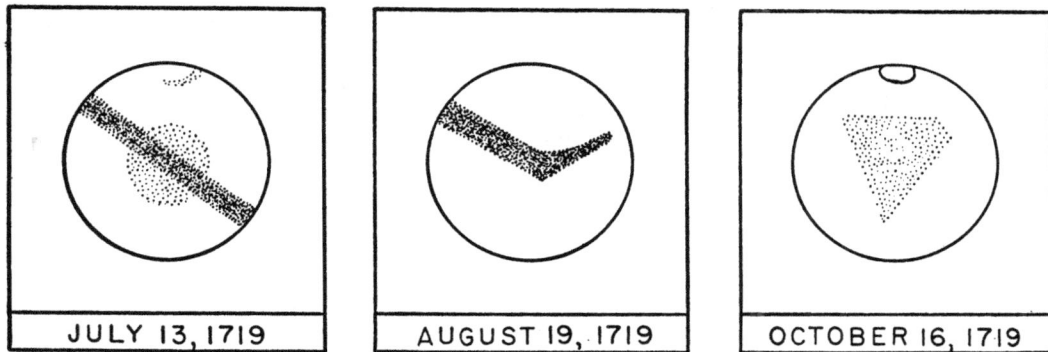

Fig. 4. Three drawings of Mars by Giacomo Filippo Maraldi

server was William Herschel, and the instruments were the new telescopes which he himself had built and which surpassed in power anything ever built before.

Herschel's main interest in making the observations is described in the titles of the papers he published later. The first one was called "Astronomical observations on the rotation of the planets round their axes," and the second, "On the remarkable appearances at the polar regions of the planet Mars, the inclination of its axis, the position of its poles, and its spheroidical figure; with a few hints relating to its real diameter and atmosphere." They were published in the *Philosophical Transactions* for 1781 and 1784, respectively. The drawings accompanying these papers mostly look like mere scribbles; Herschel was an excellent observer but a poor draftsman. His observations convinced him that the white polar spots—he also noted their eccentricity to the geographical poles—so carefully and noncommittally labeled by Maraldi were actually accumulations of snow and ice. He stated unhesitatingly that Mars has an atmosphere, "so that its inhabitants probably enjoy conditions analogous to ours in several respects." He was the first to observe color changes on Mars. He established when spring begins on Mars. He found the axis of Mars inclined to the ecliptic by 59 degrees, 42 minutes, and calculated the length of the Martian day as 24 hours, 39 minutes, 21.67 seconds.

Herschel's papers on Mars mark the last important work done prior to the nineteenth century. The body of knowledge about the fourth planet from the sun was not yet very large, but one could hope for improved equipment which would make it possible to learn more. This hope, as we know in retrospect, was fully justified. But nobody could guess what big surprises the nineteenth century was to bring.

2. "DE MOTIBUS STELLAE MARTIS"

T IS amusing—though admittedly not very useful—to speculate on what might have happened if Kepler had not worked "on the motions of Mars" when he set out to establish the true shape of a planetary orbit. If he had chosen the observations of the planet Venus instead of those of Mars everything might have been quite different. After reducing all the measurements of apparent positions in the sky to true positions in space, making due allowance for minor observational errors, Kepler might have found himself compelled to conclude that the orbit of Venus *is* a circle. And after that he would have had to make a very difficult decision: either all the observations of all the other planets had always been very much in error, or else it had to be accepted as a fact of Nature that one planet had one type of orbit and another planet a fundamentally different type.

To us, just because of Kepler's long and finally successful struggle with multitudes of figures, the fundamental mechanics of the solar system are rather simple. Of course the orbits of all planets are ellipses. It just so happens that some orbits are more eccentric than others, and that of Venus happens to be very nearly circular. The deviations of Venus's orbit from a true mathematical circle are so small that measurements in pretelescopic times could not show them clearly—not even Tycho Brahe's large quadrants. And anybody would have been fully justified in ascribing to observational error such deviations as might have appeared. If Kepler had picked Venus he very likely would not have found either the first or the second of his laws. Their discovery might have been delayed by as much as a century and they would now bear somebody else's name. Kepler always thought of himself as a man who was not particularly pursued by good luck, and in general he was correct in thinking so. But his choice of the orbit of Mars with its pronounced eccentricity and the noticeable changes in orbital velocity which go with a high eccentricity certainly was a case of good luck. Of the inner planets only Mercury has an equally eccentric orbit, or rather an even more eccentric one, but fast-moving Mercury, because of its proximity to the sun, is hard to observe even with optical instruments, which were still lacking in Tycho Brahe's time.

Before we can go on with the story of the planet Mars it is necessary to talk about planetary motions in general. The planet Mercury is the one closest to the sun; then follow Venus, earth, Mars, the minor planets (planetoids or asteroids are alternate names for them), Jupiter, Saturn, Uranus, and Neptune. Pluto, sometimes a little closer to the sun than Neptune but on the average farther away, is the outermost known member of our solar system. All these planets move around the sun in the same direction. Each has an orbital velocity of its own, but in the direction of movement there is agreement. It is counterclockwise when the solar system is viewed from "above"—meaning from a point very far above the northern hemisphere of the earth. The planetary orbits also share very nearly the same plane. If, as is usually done, one takes the orbital plane of earth, the ecliptic, as the reference, the other planets may be said to move very nearly in the ecliptic, with only minor deviations. The whole solar system, therefore, would be nearly as flat as the rings of Saturn if it were not for a number of culprits with tilted orbits along which they can rise high above the ecliptic at certain times and descend far below it at other times. The two chief culprits in that respect are the two planets in the extreme positions in the solar system, the innermost and the outermost, Mercury and Pluto. Some more strongly tilted orbits can be found among the planetoids, but all kinds of unusual behavior are customary with them.

But while the planets share the direction of movement, and the majority of them move fairly much in the same plane, there is no rule that governs the positions of the major axes of their orbits. The major axis of an ellipse is the axis drawn through the two focal points; the other one, the minor axis, is drawn at a right angle to the major axis through the center of the ellipse. The two ends of the major axis mark two points of great importance, the points where the planet is either nearest the sun (at one end of the major axis) or farthest from it (at the other end). The point nearest the sun is technically known as the perihelion, from Greek *peri*, meaning "around," and *helios*, the sun. The point farthest from the sun is named aphelion. If the solar system were to be designed today by a mathematician (or a drill sergeant), he would, no doubt, see to it that all the major axes of the planetary ellipses coincided, so that all aphelia were lined up on one side and all perihelia on the opposite side. Reality lacks such military precision. It would be possible, even

21

though it does not happen to be the case in our solar system, for the perihelion of one planet to lie in the same direction from the sun as the aphelion of another planet. Figure 5 shows the perihelia of both Mercury and Mars to illustrate this point. It also shows why, in the preceding chapter, some of the past oppositions of

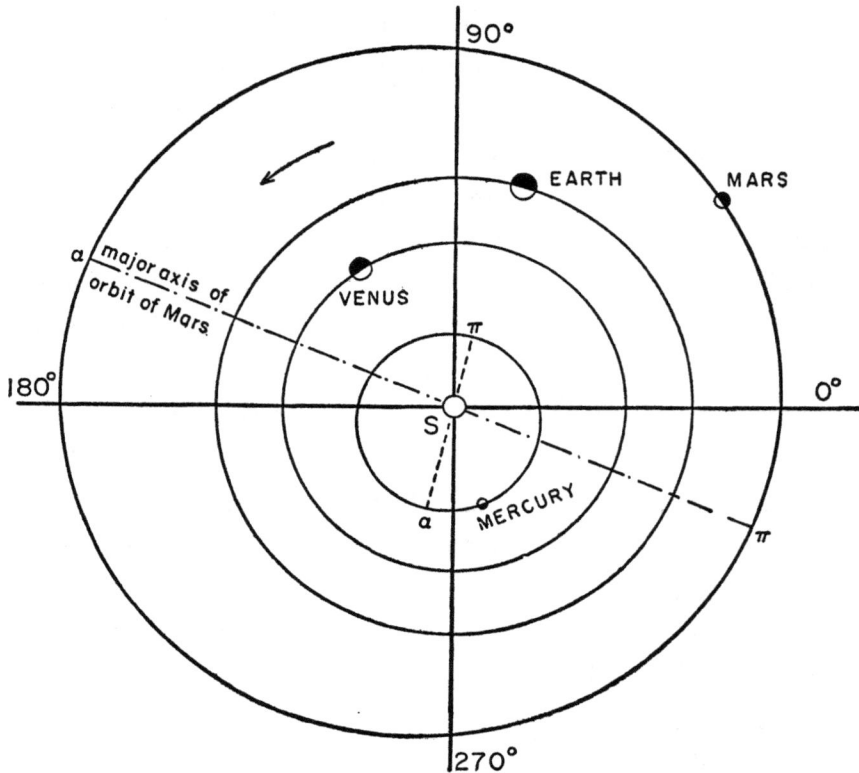

FIG. 5. Orbits of the inner planets of the solar system, drawn to scale. While the orbits of Venus and of earth are very nearly circular, those of Mercury and of Mars show pronounced eccentricity. The perihelion points of both the latter planets are indicated by the Greek letter pi (π) and their aphelion points by the Greek letter alpha (α).

Mars have been referred to as "good" oppositions, with the implication that not all oppositions are of equal value.

Since the term "opposition" means that the sun, the earth, and another planet, in this case Mars, form a straight line with the earth in the center, we only have to imagine that one opposition takes place along the line marked 0° on Fig. 5 and a later opposition takes place along the line marked 180°. Obviously an observer on earth looking at Mars at an 0-degree opposition has to look across a shorter distance than does an observer looking at Mars during a 180-degree opposition. The most favorable case, of course, would be if earth happened to meet Mars along the line which is the major axis of the Martian orbit, because then Mars would be in perihelion. As far as simple distance is concerned—there are a few other criteria—a perihelion opposition is better than any other opposition elsewhere along the orbit could possibly be. Conversely, as far as distance is concerned, an aphelion opposition is the least favorable opposition possible. Of the more recent oppositions, the one of 1924 came close to being a perihelion opposition, while the one of 1933 was almost precisely an aphelion opposition.

The terms perihelion and aphelion, when used in connection with oppositions of Mars, refer to the perihelion and aphelion of Mars and not those of earth. The reason for this usage is that it is the position of Mars on its orbit which counts. Its perihelion distance from the sun is 128 million miles and its aphelion distance 154 million miles. In the case of the earth the difference between perihelion and aphelion is "only" 3 million miles. But even this is helpful to some extent in reducing the distance between earth and Mars at a perihelion opposition, because the earth's aphelion happens to lie in the same general direction from the sun—one might say in the same sector—as the perihelion of Mars. Earth passes aphelion 94.5 million miles from the sun on July 4 and crosses the major axis of the Martian orbit (and with it the direction of Mars' perihelion) on August 28. The earth, on that date, is still farther from the sun than its average, and that helps. What is gained in the earth-at-aphelion and Mars-at-perihelion sector of the two orbits is necessarily lost on the other side. Earth passes its perihelion 91.5 million miles from the sun on January 4 and crosses the major axis of the Martian orbit (and this time the direction of Mars' aphelion) on February 23. Mars is then at its farthest from the sun while the earth is even nearer the sun than its average distance.

Supposing now that an opposition took place at o degree in the year X, how long do we have to wait for the next opposition? The principle is easy to understand. The earth moves faster than Mars and has completed its circle around the sun and returned to o degree after one year, or 365.25 days. Mars, meanwhile, has fallen far behind; it will not reach o degree on its orbit again until after 687 earth days have passed. This is 321.75 days more than an earth year, so that the earth has made most of a second circuit around the sun by the time Mars reaches o degree. When earth gets to this point, 43.5 days later, Mars has moved too and is still a little ahead; earth has not quite caught up. To catch the slower-moving Mars needs another 49.5 days. Since 2 earth years are 730.5 days, the whole process requires

730.5 + 49.5 = 780 days, or 2 years and 7 weeks. Naturally the opposition following the one of the year X does not take place along the o-degree line but about 49 degrees beyond.[1]

This, however, is not yet the whole story. As has been mentioned in Chapter 1 the figure of 780 days does not hold true for the time interval between two actual successive oppositions. Since Nature is always right, our explanation must be at the very least incomplete, and so it is. So far it has been tacitly assumed that both planets travel along their respective orbits with uniform velocities. As we well know, they don't, and what happens is that during an aphelion opposition Mars travels considerably more slowly than the averaged figure used. Earth, for its part, is in the perihelion sector of its orbit and traveling somewhat faster than the averaged figure. Consequently the earth will catch up with Mars faster when the two planets are on that sector of their orbits. But when Mars is in its perihelion sector and especially fast, earth is in its aphelion sector and especially slow, so that it takes an extra week or so until the earth can catch up with Mars.

And this brings up another question. Since the opposition that follows the opposition of the year X does not take place on the o-degree line, how many oppositions, or how many years, will have to go by until another opposition takes place on the o-degree line? To answer this question it is best to express the Martian year in terms of earth years and to say that Mars returns to a specific point on its orbit after 1.8808 terrestrial years. Eight revolutions of Mars take, therefore, about 15 earth years; the trouble is that it is "about" and not precisely 15 years. There is a difference of 0.046 earth years. Expressed in days 0.046 years means very nearly 17 days—the precise figure is 16 days and 19.2 hours—and the two planets would not form a straight line with the sun. Instead they would form an angle of a little more than 16 degrees, which is very considerable. But this discrepancy grows smaller the longer the time interval under consideration, as can be seen from the table on page 25, which was worked out by Robert Henseling more than 30 years ago.

1. The simplified calculation on which these figures are based goes like this: the earth requires 365.256 days to move through the full circle of 360 degrees, hence the earth moves not quite 1 degree per day. The figure is found by dividing 360 by 365.256, with the result of 59 minutes and 48.2 seconds of arc. Mars needs 686.98 days for a full circle. The same kind of division results in the figure of 31 minutes and 26.5 seconds of arc per day. Expressed in seconds of arc, earth's movement is 3548.2. Subtracting Mars' movement of 1886.5 seconds of arc shows a daily advance of earth over Mars of 1661.7 seconds of arc. A full circle contains 1,296,000 seconds of arc and this figure divided by the daily advance gives the result of 779.92 days.

NUMBER OF COMPLETE REVOLUTIONS OF MARS	CORRESPONDING NUMBER OF TERRESTRIAL YEARS	DISCREPANCY (DEGREES)
8	15.046	16.6
17	31.974	9.4
25	47.020	7.2
42	78.994	2.2
151	284.0008	0.3

One can say, therefore, that any given Mars opposition will repeat itself almost precisely after 284 years. The fine opposition of 1719 which was utilized by Maraldi will be repeated to all practical intents and purposes in 2003. The less favorable but still good opposition of 1666 of which Cassini took advantage was repeated in 1950. The spectacularly good opposition of 1924 was a repetition of the one of 1640, which came too soon after the invention of the telescope to do any good. If a difference of 2 degrees—about four times the diameter of the full moon—should be considered acceptable, almost-repetitions of a given opposition take place after 79 years have elapsed.

With the aid of this 284-year rule and a list of previous oppositions one can predict the next opposition with fair accuracy. In reality every opposition is of course calculated separately. And whenever a perihelion opposition is forthcoming, the planetary observers among the astronomers prepare to do some new work on Mars' southern hemisphere. The reason it is the southern hemisphere can be seen in Fig. 8. The plane of the orbit of Mars does not coincide with the ecliptic, although the difference is small, amounting to a trifle less than 2 degrees. At the end of August—the drawing has been made to show a perihelion opposition—the earth is about 3 million miles "above" (that is, to the north of) the plane of the orbit of Mars.

During a perihelion opposition the northern hemisphere of earth is having summer. But on Mars, because the axis of the planet points in a different direction, the northern hemisphere has winter and the north pole is in darkness. The south pole of Mars enjoys a long, long day, and the southern hemisphere in general has summer. Hence the southern hemisphere of Mars is visible, although one could wish that we had a few big telescopes in our southern hemisphere for observations, because for the northern observatories Mars is not very high above the horizon. The picture is, of course, reversed during an aphelion opposition. Then the northern hemisphere of earth is having winter, and the northern hemisphere of Mars, summer. Moreover, Mars is high in the sky for the northern observatories of earth, so that

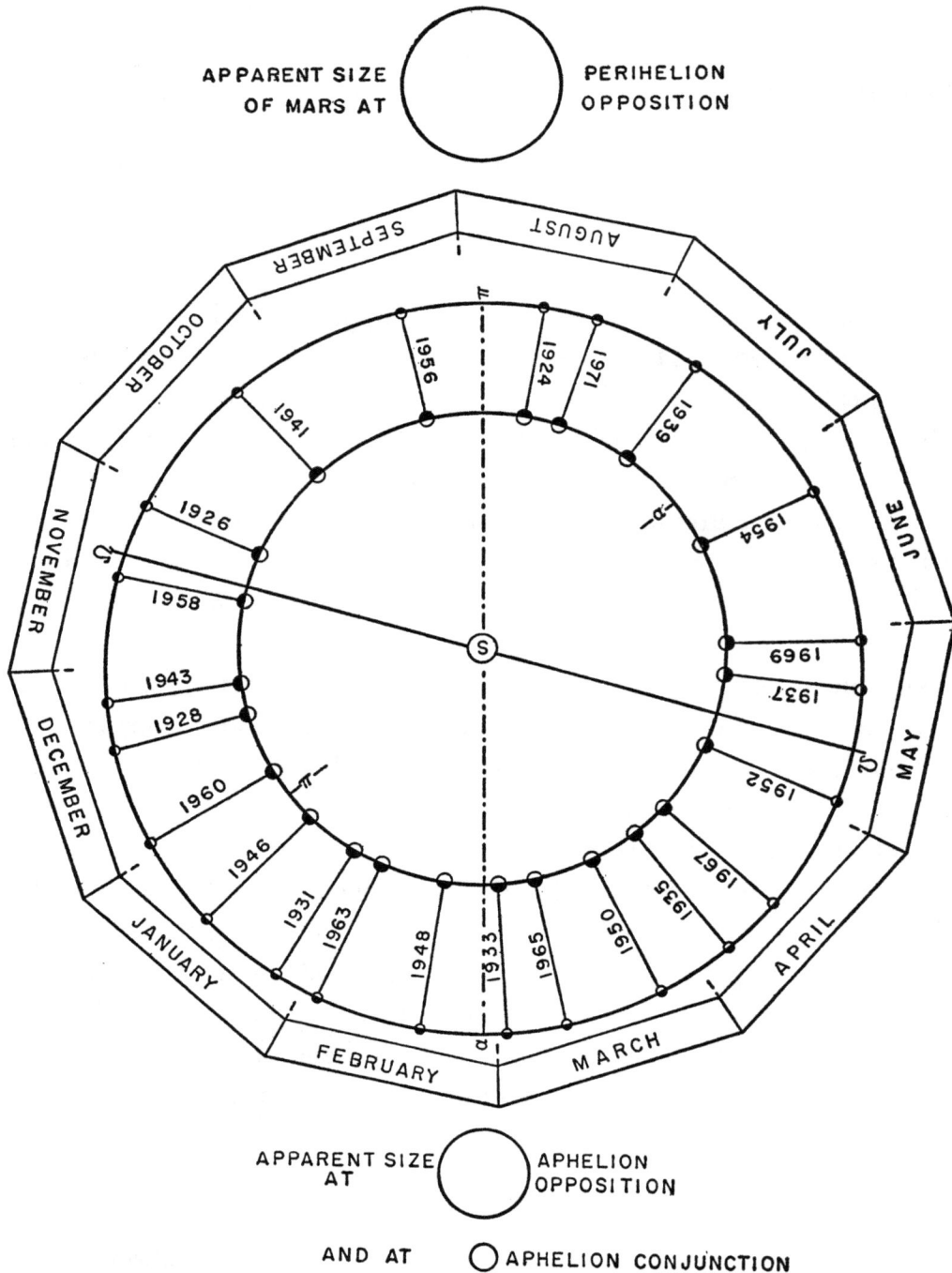

FIG. 6. Oppositions of Mars from 1924 to 1971, surrounded by a calendar to show the time of year in which they occurred or will occur. The broken line is the major axis of the orbit of Mars, connecting aphelion and perihelion. The aphelion and perihelion of earth are indicated on earth's orbit. The solid line is the line connecting the nodes; if the plane of the orbit of Mars is identical with the plane of the printed page the portion of the earth's orbit above that line should be imagined to rise slightly out of the page, while the portion of the earth's orbit below that line should be imagined to be slanting a little below the page.

"DE MOTIBUS STELLAE MARTIS"

THE PLANET MARS

	MILES	KILOMETERS
Distance from sun		
Mean (1.5237 A.U.*)	141,500,000	228,000,000
Aphelion (1.6658 A.U.)	154,100,000	248,000,000
Perihelion (1.3826 A.U.)	128,000,000	206,000,000
Distance from earth		
Perihelion opposition	34,797,000	56,000,000
Aphelion opposition	61,516,000	99,000,000
Aphelion conjunction	340,000,000	547,200,000
Orbital velocity per second		
Mean	14.98	24.11
At aphelion	13.64	21.95
At perihelion	16.45	26.37
Escape (parabolic) velocity, per second	3.13	5.04
Circular velocity at surface, per second	2.21	3.56
Equatorial diameter	4220	6780

Length of day
Sidereal   24 hours, 37 minutes, 22.668 seconds
Solar   24 hours, 39 minutes, 35.247 seconds

Length of year (668.599 Mars days)	686.979 earth days
Eccentricity of orbit	0.09336
Mean sidereal motion in 24 hours	1886.519 seconds of arc
Inclination of orbit to ecliptic	$1° 50' 59.8''$
Inclination of Martian equator to its orbit	$25° 10'$
Heliocentric longitude of node (1956)	$49° 13' 05.5''$
Heliocentric longitude of perihelion (1956)	$335° 14' 56.6''$
Mass (earth = 1)	0.108
Volume (earth = 1)	0.151
Density (earth = 1)	0.710
Density (water = 1)	3.910
Surface area (earth = 1)	0.278
Gravity at surface (earth = 1)	0.38

* A.U. stands for "astronomical unit," the distance of the earth from the sun.

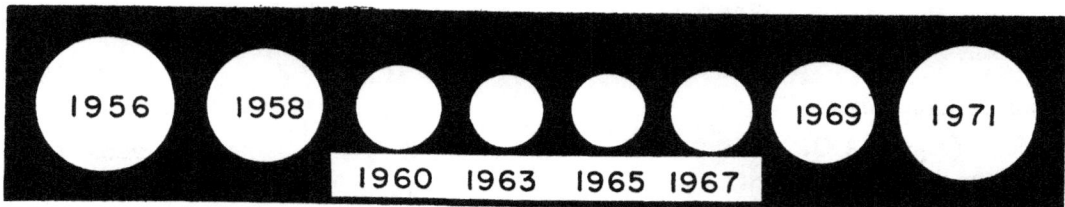

FIG. 7. The apparent size of the disk of Mars as it will appear in a telescope of a given magnification during the oppositions from 1956 to 1971.

LIST OF OPPOSITIONS OF MARS, 1939–1975

YEAR	DATE	SHORTEST DISTANCE FROM EARTH		LARGEST APPARENT DIAMETER (SEC. OF ARC)
		MILES	KILOMETERS	
1939	July 27	36,171,000	58,212,000	23.9
1941	Oct. 3	38,508,000	61,983,000	22.7
1943	Nov. 28	50,599,000	81,431,000	17.3
1946	Jan. 9	59,800,000	96,239,000	14.6
1948	Feb. 18	63,000,000	101,389,000	13.8
1950	March 24	60,700,000	97,687,000	14.4
1952	May 2	52,400,000	84,330,000	16.6
1954	June 25	40,300,000	64,857,000	21.8
1956	Sept. 11	35,400,000	56,971,000	24.7
1958	Nov. 16	45,100,000	72,580,000	19.2
1960	Dec. 29	56,200,000	90,445,000	15.4
1963	Feb. 3	61,800,000	99,457,000	14.0
1965	March 8	61,800,000	99,457,000	14.0
1967	April 13	56,200,000	90,445,000	15.4
1969	May 29	45,300,000	72,900,000	19.1
1971	Aug. 6	34,600,000	55,683,000	25.0
1973	Oct. 21	40,600,000	65,340,000	21.6
1975	Dec. 13	53,100,000	85,450,000	16.5

an aphelion opposition has a number of ameliorating features in spite of the greater distance between the planets.

One more point ought to be mentioned, although it is not too important as far as observational practice is concerned. If the orbits of Mars and of the earth were in precisely the same plane, the moment where the centers of the sun, of the earth, and of Mars form a straight line—the moment of opposition—would also be the instant of closest approach. But in reality the orbits are tilted against each other, and the motion of the planets may be compared to cars moving on slanting ramps; therefore the moment where they are closest to each other is not necessarily the same as the moment where they form a straight line with a third object. In other words, minimum distance does not take place at the same instant as opposition, although the time interval separating the two is rather small. During the opposition of 1877 minimum distance was attained on September 2; opposition took place on September 5. In 1881 minimum distance was attained on January 30; opposition took place on January 31. In 1888 the date of the opposition was April 11, and the date of minimum distance April 17, which is almost the longest interval possible. In 1924, however, both events occurred on August 23, but minimum distance preceded opposition by 18 hours. Only if an opposition took place while both planets were passing through the line of the nodes (see Fig. 6) could minimum distance and opposition occur at the same instant, because in that case both planets are equally far north or south.

The seasons on a planet with an orbit reasonably resembling a circle are in the main caused by the position of the planet's axis. On earth the northern hemisphere has summer while the planet is at aphelion: the fact that the sun is higher over the northern hemisphere because of the earth's axial tilt is far more important than the additional distance of 3 million miles. Similarly, the fact that the sun is lower over the horizon in January is more important than the fact that the earth is passing perihelion. The southern hemisphere, of course, has summer at perihelion and winter at aphelion, and some climatologists have tried to show that earth's seasons are somewhat more pronounced on the southern hemisphere, that the summers are hotter and the winters colder. However, the case is by no means clearcut; the statistics might just as easily be influenced by local factors. The 3 million miles between earth's aphelion and perihelion do not seem to matter much.

With Mars the story is different, for there the distance from the sun varies by 27 million miles in the course of a Martian year. The hemisphere which has summer

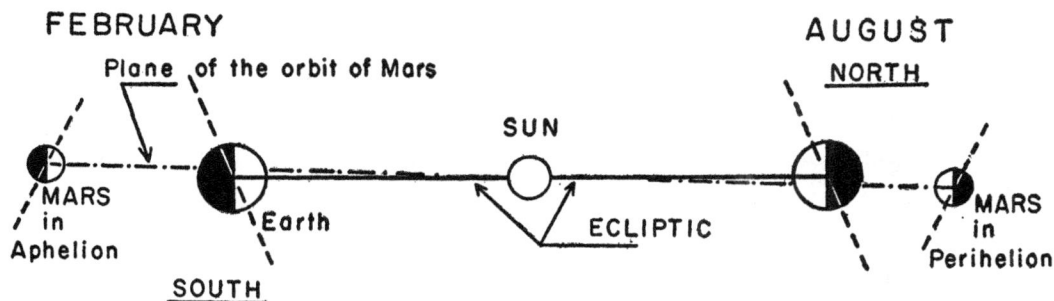

FIG. 8. The inclination of the orbital plane of Mars to that of earth, the ecliptic, and the position of the axes of both planets. If earth meets Mars during the northern summer on earth, it must be at a time when the southern hemisphere of Mars has summer. Conversely, during the northern winter on earth the northern hemisphere of Mars has summer.

at perihelion must have a hotter summer than the one that has it at aphelion. Conversely the hemisphere with "aphelion winter" must have a colder winter than the other. And because of the noticeable difference in orbital velocity the "perihelion summer" must be short and the "aphelion winter" must be long. (The southern hemisphere of Mars is the one that has this more extreme climate.) For the same reason the Martian seasons differ in length to a much greater degree than do the seasons on earth. On earth the northern summer lasts 93½ days, the southern summer 89 days. On Mars the northern summer, the aphelion summer, lasts 182 (earth) days, while the southern, perihelion summer lasts only 160 (earth) days. Because the Martian day is some 37 minutes longer than the earth day, the Martian year contains 668.6 Martian days. The lengths of the seasons on the two planets are as follows:

SEASONS			
NORTHERN HEMISPHERES	SOUTHERN HEMISPHERES	MARS (MARTIAN DAYS)	EARTH (EARTH DAYS)
Spring	Fall	194.2	92.9
Summer	Winter	176.8	93.6
Fall	Spring	141.8	89.7
Winter	Summer	155.8	89.1
		668.6	365.3

On both planets, then, the southern summers are shorter and the southern winters longer than their northern counterparts. But on Mars the difference is far more pronounced, and of course each season lasts almost twice as long as it does on earth. Because of the different position of the Martian axis the seasons look reversed; the northern hemisphere of Mars has summer when earth passes it with winter on its northern hemisphere. Neither earth nor Mars has a bright star in the southern sky to mark the position of its pole. But Mars lacks a northern pole star, too; its north pole happens to be in an area which to the unaided eye is "empty." The bright star nearest to the northern celestial pole of Mars is Deneb, but Deneb is a full 10 degrees from that point. Naturally, the starry sky over a Martian landscape does not differ from the sky we see at night; all the fixed stars maintain the same relative positions that they have for us.

Still, there are differences. At the proper intervals an enormous morning or evening star hangs over the Martian landscape, accompanied by a smaller star less than one-tenth its brilliance. These are the earth and its moon. Then there is a lesser

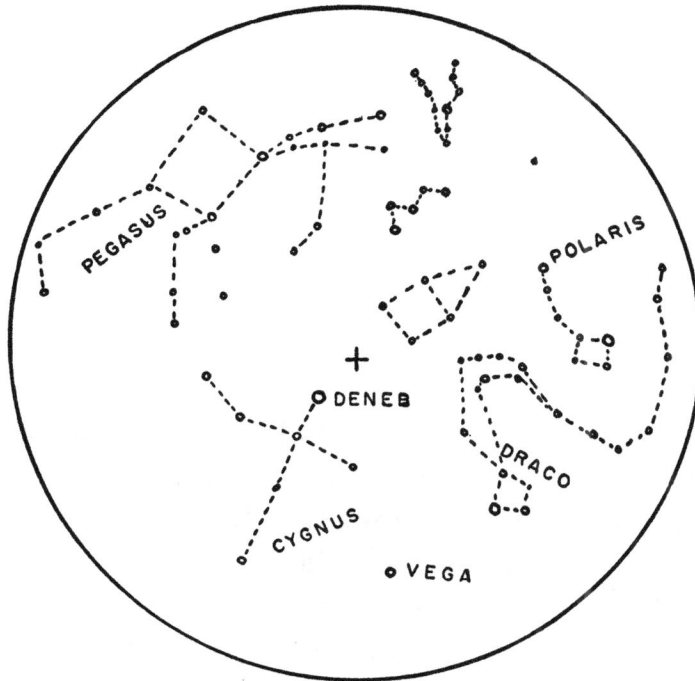

FIG. 9. While the celestial north pole of earth is marked by the close proximity of the star Polaris, the celestial north pole of Mars happens to lie in an area of the sky that is devoid of bright stars. In the sketch it is marked by the cross near Deneb.

evening or morning star which appears more often but is not so bright—Venus. Jupiter and Saturn must be much brighter at times than they appear from the earth. Occasionally some of the larger planetoids will be temporary and faint naked-eye stars in the Martian sky. And there are also Mars' two moons. They have a story of their own, which is told in Chapter 3.

3. 1877

IT IS an interesting and rarely mentioned fact that at the time the older Herschel made his observations of Mars—from 1777 to 1783—astronomy was the most highly developed of all the natural sciences. Herschel himself had contributed heavily to this advanced status, but even without him astronomy would have been far ahead of the others. Chemistry in the modern sense did not yet exist; it was still ruled by the phlogiston theory, a concept now forgotten by everybody except historians. Chemists did know how to make certain substances, mostly of pharmacological value, but that was about the sum total of their knowledge. Electricity still had to be discovered, though a few electrical phenomena were known. Zoology and botany were still in the state of collecting specimens and information, without much understanding of either natural relationships or biological functions. Meteorology was a thing of the future, and geology likewise. So was metallurgy. Medicine fumbled, and surgery was in the stage of benevolent torture, with occasional successes. Geography was doing relatively well, but still had enormous gaps, comprising most of the interior of Africa and virtually everything to the south of the 60th parallel of southern latitude. Physics was considerably less developed than astronomy, but still ahead of the other natural sciences. The result was that for about three-quarters of a century the other natural sciences seemed to progress faster than astronomy did. Though astronomy itself made startling advances, the advances made by the other sciences, which were just getting under way, were even more startling.

As regards the planet Mars, William Herschel's work was considered for many years as a "milestone." It was not until the discoveries of the year 1877 took place that a newer "milestone" was established. This, however, must not be taken to mean that no useful work on Mars was done during the interval between Herschel and the year 1877. In Herschel's own time another patient and careful observer had gone to work on the red planet. He was Dr. Johann Hieronymus Schroeter, who spent a large portion of an inherited fortune to build himself a then up-to-date observatory at Lilienthal near Bremen. This observatory was destroyed during the Napoleonic wars, presumably because Schroeter, in daylight hours, was an official of an "enemy principality." Schroeter's observations covered the oppositions from 1785 to 1802, but they were not printed until long after the author's death. The manuscript and drawings were kept in the library of the Observatory of the University of Leyden and were finally published by H. G. van de Sande Bakhuyzen, the observatory's director. The year of publication was 1881, and considering the important work that had been done during the few preceding years and the work still in progress at the time, the publication was merely a kind gesture to call attention to Schroeter's work. But it is interesting to note that Schroeter is apparently the inventor of the term "areography" for the description of the Martian surface. (His title was *Areographic Contributions . . .* , etc.) The term was coined in analogy to "geography," using *Ares*, the Greek word for Mars, instead of the Greek word for the earth (*gaia*).

Another private observatory entered the story of Mars during that interval between Herschel and Schroeter and the year 1877. It was located in the Tiergarten in Berlin and had been built by an enthusiastic amateur astronomer, the banker Wilhelm Beer, a brother of the composer Giacomo Meyerbeer. There Wilhelm Beer, in collaboration with J. H. von Mädler, produced a detailed map of the moon (1834) which for many years was the last word in lunar matters and a greatly admired masterpiece. At that private observatory the first map of Mars was also drawn (see Plate III *a*). It was decidedly inferior to the work on the moon, but was the earliest attempt to put all areographic information into one picture.

Of course every important astronomer of the period, and a large number of unimportant ones, did some work on Mars when an opposition came around. The more valuable studies are linked with half a dozen names, among them Sir Joseph Norman Lockyer in England, Father Pietro Angelo Secchi in Italy, Dominique François Arago in France, and F. Kaiser in Holland. During the opposition of 1864, which took place late in November and was therefore not a first-class perihelion opposition but still good enough, the Reverend W. R. Dawes (working at Hopefield Observatory, Haddenham, Buckinghamshire, England) made a number of exceptionally fine drawings of Mars, two of which are reproduced as Plate III *b*. A few years later another English astronomer, Richard Anthony Proctor, published another map of Mars, the first map to contain names for the various surface features (see Plate ɪv *a*). Surprisingly this map caused a storm of criticism, and not even because it wasn't a very good map. The critics were an international lot, and they all felt annoyed because Proctor had not been fair in assigning the names. In fact, they said, he had been almost chauvinistic—all the important "continents" and "seas" bore the names of English astronomers!

If one supplies the map of a planet with the very necessary network of degrees of longitude and latitude, there is one place where arbitrary action is required. The poles, and with them the equator, are given by the rotation of the planet. The arctic and antarctic circles are given by the axial tilt of the planet, and so are the two tropics. But the zero meridian is arbitrary. On earth we use the location of the old Observatory of Greenwich in England as the zero meridian—for quite some time the Observatory of Paris was a strong competitor. For Mars, Proctor used the location of a feature which he named Dawes Forked Bay. The location of the zero meridian is the only feature of Proctor's map that has survived to this day.

At the time of the publication of Proctor's map the positions of earth and of Mars along their orbits had become such that no close opposition was to be expected for some time. The opposition of 1867 was in early January; the one of 1869 almost an aphelion opposition in middle February; the one of 1871 in late March not much better (after Martian aphelion). In 1873 the opposition fell in early May and the one of 1875 in late June. The mutual positions of the planets were building up to an opposition just after Martian perihelion in 1877.

The first of the two big discoveries made during that opposition was announced from Washington, D. C., and originated with Asaph Hall, professor of mathematics at the Naval Observatory. After those nights of the latter part of August, 1877, Mars was no longer moonless, as the books had been saying ever since the time of Cassini. Professor Hall stated that he had found that Mars had two tiny moons. The announcement must have been read with an audible gasp by many people. Strangely enough, the "two moons of Mars" had quite a history outside of professional astronomical literature. Johannes Kepler is said to have remarked on one occasion that Mars might have two moons, and in speculative philosophy they must have been mentioned quite often. François Marie Arouet, better known as Voltaire, in his satirical tale "Micromégas," about the voyage of a giant from Sirius who visits the solar system, has a passage about them:

On leaving Jupiter our travelers crossed a space of about a hundred million leagues and reached the planet Mars. They saw two moons which wait on this planet, and which

34

have escaped the gaze of astronomers. I know well that l'abbé Castrel wrote against the existence of these two moons; but I agree with those who reason from analogy. These good philosophers know how difficult it would be for Mars, which is so far from the sun, to get on with less than two moons.

"Micromégas" was published in 1750—thirty years after Dean Swift wrote his "Voyage to Laputa," telling the adventure of his Captain Lemuel Gulliver with the inhabitants of the flying island. The Laputan astronomers

spend the greatest part of their lives in observing the celestial bodies, which they do by the assistance of glasses far excelling ours in goodness. For this advantage hath enabled them to extend the discoveries much farther than our astronomers in Europe; for they have made a catalogue of ten thousand fixed stars, whereas the largest of ours do not contain above one-third part of that number. They have likewise discovered two lesser stars, or satellites, which revolve about Mars, whereof the innermost is distant from the centre of the primary planet exactly three of his diameters, and the outermost five; the former revolves in the space of ten hours, and the latter in twenty-one and a half; so that the squares of their periodical times are very near in the same proportion with the cubes of their distance from the centre of Mars,[1] which evidently shews them to be governed by the same law of gravitation that influences the other heavenly bodies.

These lines from the third chapter of the "Voyage to Laputa" are almost astonishing, not only because of the prediction itself, but because Dean Swift came so close to some of the actual values that can be read from Fig. 10. Possibly prompted by this literary fiction, possibly because they were (as Asaph Hall later said about himself) "tired of reading that 'Mars has no moons,'" several astronomers had actually searched for Martian satellites. William Herschel's notebooks of the year 1783 contain several entries testifying to his search, which did not bring any positive results. At a later date it became known, via the publication of correspondence with colleagues, that Professor d'Arrest, director of the Copenhagen Observatory, had searched for Martian moons during the opposition of 1864 and possibly also during the preceding and somewhat better opposition of 1862. But he was unsuccessful, and Asaph Hall in 1877 was very nearly unsuccessful too. After weeks of fruitless watching—he began his search a good distance from the planet, slowly "moving in" more and more closely—he was ready to give up. It was only at the insistence of his wife, who also acted as his secretary, that he went back to the telescope for another try. Her argument was that an equally favorable opposition would not come soon again,

1. Dean Swift was quoting Kepler's Third Law here.

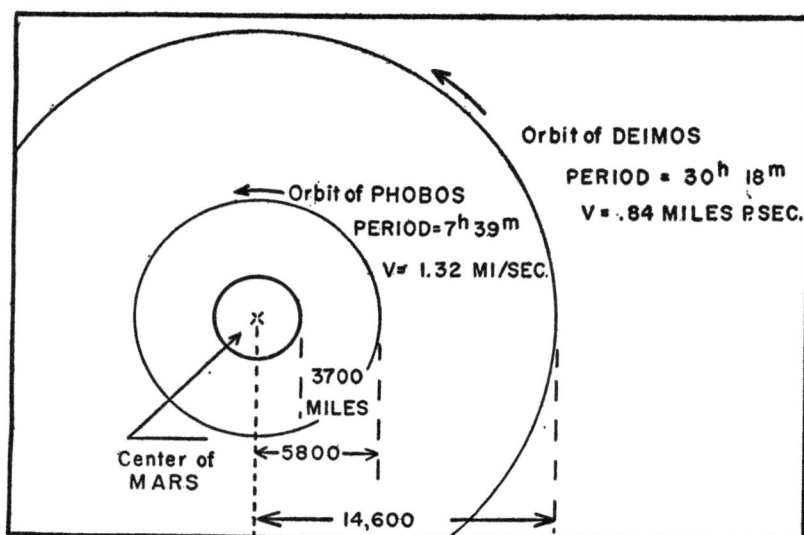

FIG. 10. The orbits of the two moons of Mars, drawn to scale. The diameter of Phobos is about 10 miles, that of Deimos not more than 5 miles.

and that at least the negative result should be established as well as a negative fact can be.

What happened then is best told in Asaph Hall's own words:

My search for a satellite was begun early in August. . . . At first, my attention was directed to faint objects at some distance from the planet; but all these proving to be fixed stars, on August 10 I began to examine the region close to the planet, and within the glare of light that surrounded it. This was done by sliding the eye-piece so as to keep the planet just outside the field of view, and then turning the eye-piece in order to pass completely around the planet. On this night I found nothing. The image of the planet was very blazing and unsteady, and the satellites being at that time near the planet, I did not see them. The sweep around the planet was repeated several times on the night of the 11th, and at half past two o'clock I found a faint object on the following side and a little north of the planet, which afterward proved to be the outer satellite. I had hardly time to secure an observation of its position when fog from the Potomac River stopped the work. Cloudy weather intervened for several days. On the night of August 15, the sky cleared up at eleven o'clock and the search was resumed; but the atmosphere was in a very bad condition, and nothing was seen of the object, which we now know was at that time so near the planet as to be invisible. On August 16 the object was found again on the following side of the planet, and the observations of that night showed that it was moving with the planet, and, if a satellite, was near one of its elongations. On August 17, while waiting and watching for the outer satellite, I discovered the inner one. The observations of the 17th and 18th put beyond doubt the character of these objects and

36

the discovery was publicly announced by Admiral Rodgers. Still, for several days the inner moon was a puzzle. It would appear on different sides of the planet in the same night, and at first I thought there were two or three inner moons, since it seemed to me at that time very improbable that a satellite should revolve around its primary in less time than that in which the primary rotates. To decide this point I watched this moon throughout the nights of August 20 and 21 and saw that there was in fact but one inner moon, which made its revolution around the primary in less than one-third the time of the primary's rotation, a case unique in our solar system.

Of the various names that have been proposed for these satellites I have chosen those suggested by Mr. Madan of Eton, England, viz:

DEIMOS for the outer satellite;
PHOBOS for the inner satellite.[2]

At first the diameters of these moons were guessed to be 30 miles for Phobos and 20 miles for Deimos. This has since been revised downward to 10 and 5 miles, respectively. The discovery of these two tiny moons gave rise to a whole host of questions, all of them somehow interlinked. As Asaph Hall had pointed out at once, the case of a satellite revolving faster than its planet turns on its axis is unique in the solar system. Of course the satellite needs such a high speed to stay in its orbit, being as close to the planet as it is. But did this fact not hint very strongly at the possibility that these moons were really bodies from the nearby asteroid belt which had been "captured" by Mars? If this were true, wasn't it possible that the searches of Herschel and of d'Arrest had failed because Mars had actually been moonless at that time? And if this possibility was granted, it could even be stated when Mars had captured its two moons—in the time interval between d'Arrest's search of 1864 and Hall's discovery in 1877.

These were intriguing ideas, but one fact did not fit the hypothesis. If the two moons of Mars were captured asteroids, one would expect them to move around Mars in orbits unrelated to the motions of Mars itself. One of them might have an orbit like that proposed for an artificial space station of earth, with its plane vertical to the plane of the orbit of Mars. The other one might have an orbit with a plane nearly coinciding with the plane of the orbit of Mars. In reality the orbits of both moons are very nearly in the same plane and also very nearly in the plane of the Martian equator. For captured asteroids this would be a bit too much of a coinci-

2. Professor Hall remarked that these are normally the names of the horses that drew the chariot of Mars, but Mr. Madan was referring to the fifteenth book of the *Iliad:*

"He [Ares] spake, and summoned Fear and Flight to
    yoke his steeds, and put his glorious armor on,"
where Phobos (Fear) and Deimos (Flight, better Terror or Panic) appear as the attendants of Mars.

dence. Soon after the idea of captured asteroids had been voiced, the director of the U.S. Naval Observatory, Professor Simon Newcomb, wrote a letter to the British journal *Observatory* explaining why, in his opinion, the Martian moons had not been discovered earlier. During the favorable opposition of 1862 there had been only two or three telescopes powerful enough to see them. The following oppositions had not been good enough, and while the opposition of 1875 had been favorable as to distance, Mars had been too low on the horizon for observatories on the northern hemisphere to find such faint objects. These arguments at least made it clear that the two moonlets did not have to be recent captures. Whether they, in spite of their now very regular orbits, were captured at some time in the distant past is a question about which the last word still remains to be spoken.

To terrestrial explorers in the equatorial regions of Mars, accustomed to the stately progress of our own moon across the sky and its sedate sequence of change of phase, the spectacle presented by these two satellites would appear weird. Phobos, the nearer and larger moon, would look small, with an apparent diameter of about one-fifth of that of our moon. With respect to the center of Mars it completes more than three full revolutions during one Martian day. But since the surface of Mars is moving in the same direction as Phobos because of the diurnal rotation, the moonlet completes only a little more than two revolutions per day. This still means that it overtakes the motion of the surface and therefore appears to rise in the west and set in the east. From moonrise to moonrise 11 hours and 6 minutes will go by for an observer on the ground, but the time from moonrise to moonset will be only 4 hours and 18 minutes. Phobos's climb from the horizon to the zenith will take just 2 hours and 9 minutes, during which time the moon will noticeably increase in size, since it will be much closer to the observer when overhead as compared to its distance when at the horizon. Of course it will also change its phases while doing this. Moreover it will frequently enter the shadow of Mars and set, or rise, eclipsed.

Deimos, on the other hand, is traveling around Mars at a rate very little faster than the rate at which Mars turns. It would appear to hang almost motionless in the sky; for a stationary observer on the surface of Mars more than 60 hours would go by from moonrise to moonset. Because it is both farther away and smaller than Phobos, slow Deimos would not look like a moon but like a very bright star, say like Venus at its brightest, and the phase changes would not be visible as such, but would appear as changes in brightness. Naturally Deimos will also enter the shadow of Mars quite frequently but it will not be eclipsed quite as often as Phobos. On the

38

FIG. 11. A true solar eclipse is an impossibility on Mars. When the small moons pass between Mars and the sun, they will just be black dots seen briefly on the solar disk. (*After Lucien Rudaux.*)

average, Phobos can attain the "full moon" position only once for every three revolutions without being eclipsed in the process. Deimos will be eclipsed only twice in nine revolutions. Because of their smallness the Martian moons cannot cause a total eclipse of the sun; what they can do is to appear as small black spots on the shining disk of the sun. The somewhat frightening and oppressive spectacle of a total eclipse of the sun is therefore impossible on Mars. What does happen will look more like a "transit," such as we see from the earth on the rare occasions when Venus crosses the solar disk.

The second great discovery of 1877 was the one that was announced from Milan. Giovanni Virginio Schiaparelli reported that a strange and intriguing Martian feature was apparently much more important than earlier observers had realized. The existence of the white polar caps had been known since Huyghens, the existence of dark areas since Huyghens and Maraldi. Herschel had resolutely named the dark areas "seas," the polar caps "ice," and the remaining lighter areas "land." Schiaparelli added that in those lighter areas there were thin but clearly visible lines, connecting the dark areas with each other and with the regions around the polar caps. The great stir that followed the announcement was largely caused by the name Schiaparelli attached to this feature. It seems to be impossible to see something without trying to give it an explanation at once, even if only a partial explanation. And it seems equally impossible to read a word intended and stated to be a label without forming associations if the word used as a label suggests them. In this particular case nothing—but absolutely nothing—justified Schiaparelli in thinking, at that point, that the fine lines he had seen had to be at a lower level than the surrounding territory. Yet he considered them "grooves" in the landscape and called

39

them that, naturally using the Italian word for grooves, which is *canali,* a combination of letters most misleading to non-Italians.

Just as soon as Schiaparelli's first report appeared in print (in 1878), published by Galilei's old Accademia dei Lincei, which had meanwhile become the Reale (royal) Accademia dei Lincei, a large number of people published rapidly formed and firmly held opinions. An equally large number undoubtedly took the sensible course of quietly assimilating the new information and of waiting to see whether future oppositions would confirm the report, but as these people said little or nothing, they were not heard.

Those who made themselves heard were clearly divided into two camps. The representatives of one camp said, in effect: "I am a good and experienced observer; I had at my disposal a telescope as good as or better than the one used by Schiaparelli; I took advantage of the opposition. But I did not see any 'canals.' Hence they don't exist, for if they did, I would have seen them too. Hence Schiaparelli must be mistaken." Depending on individual temperament and inclinations, the statement would either end on that note or else would be followed by a usually unflattering guess as to the reasons for such a mistake. It is worth mentioning that this camp was composed exclusively of practicing astronomers.

Now, more than three-quarters of a century later, the answer to all this is very simply that the canali have meanwhile been seen by very many reputable observers on very many occasions. But even in 1878 the canali were not such a novelty as critics tried to make out. One of the drawings by Beer and Mädler, made in 1840, shows two of them very clearly. The drawings of the Reverend W. F. Dawes show several which can be identified with "canals" on later maps. There are "canals" on the drawings of Sir Norman Lockyer, and also on drawings made by de la Rue, by Kaiser, and by Lassell, all prior to 1877. There are a few "canals" on Proctor's map. Schiaparelli must have been well aware of this, for he never claimed to be the discoverer of the canali, even though others usually said he was. He was not even the first to use this word for streaks on Mars. Father Angelo Secchi had done so before him, though with reference to a very wide and rather short streak. What was new in Schiaparelli's observations was not that he had seen canali, but that he had seen so many of them; there are forty on the map he assembled from many partial and detail drawings of Mars he made during the opposition of 1877. The strange fact remains, however, that during this particular opposition nobody else seems to have seen any.

Prima Martis facies

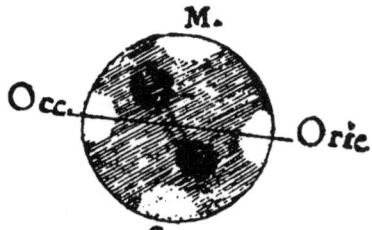

Secunda Martis facies

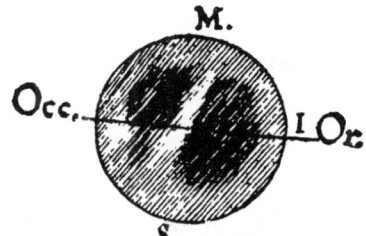

Primæ faciei
Succeſſiua conuersio

2

3

Secundæ faciei
Succeſſiua conuersio

II

III

Martis reuolutio circa axem proprium a I. D. Caſſino Teleſ: copio I. Campani obſeruata menſe Febr. Mart. April. 1666.

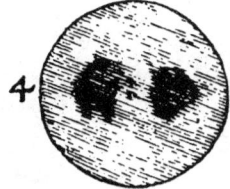
4

5

6

IV

V

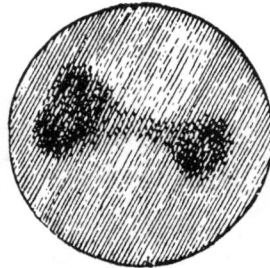
H

G

I. A series of drawings of Mars, made by Jean Dominique Cassini in the spring of 1666, which show its rotation. The drawings on the left, labeled with Arabic numerals, and the drawing labeled H show one hemisphere of the planet; the drawings on the right, labeled with Roman numerals, and the drawing labeled G show the planet's other hemisphere.

Anno 1666. Die 30. Martii hor. 2. n. s.

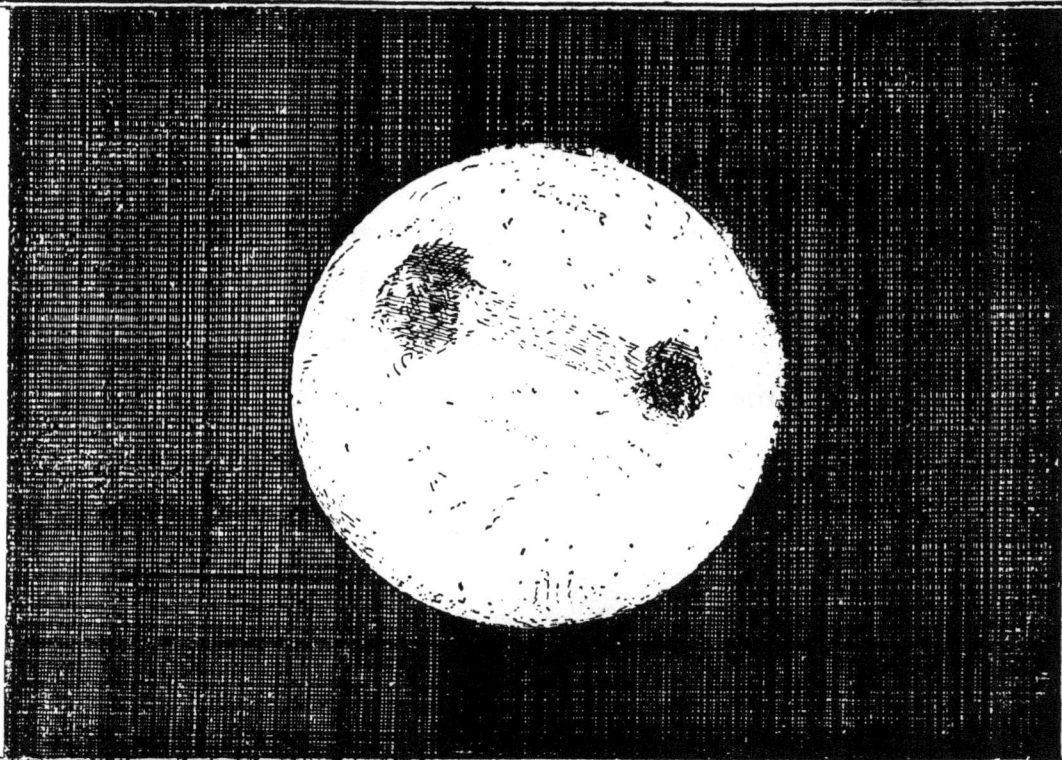

Typus Martis cum insignibus maculis Romę primùm uisis D.D. Fratribus Saluatori, ac Francisco de Serris tubo Eustachii Diuini palmorum 25, ac subinde 60. à die 24. Martii ad 30, qua die in ędibus Ill.mi D. Cesarii Giorii horá prędicta, et ipsomet Ill.mo Dño describente tub. p. 45. apparuit ut hic exprimitur inuerso modo, nigriore inter alias existente macula Orientali, pro situs obseruata uariatione eiusdem planetę circa proprium axem reuolutionis periodum indicatura, horis nempe circiter 13.

II. A drawing of Mars made by Salvatore Serra on March 30, 1666, during the same opposition utilized by Cassini. The drawing corresponds well with Cassini's drawing H and was one of the drawings by other observers which Cassini reproduced.

42

III *a*. Map of Mars drawn by Wilhelm Beer and J. H. von Mädler in 1830 and 1832. Though Beer and Mädler were famous as lunar observers, their map of Mars bears little resemblance to those drawn by other astronomers.

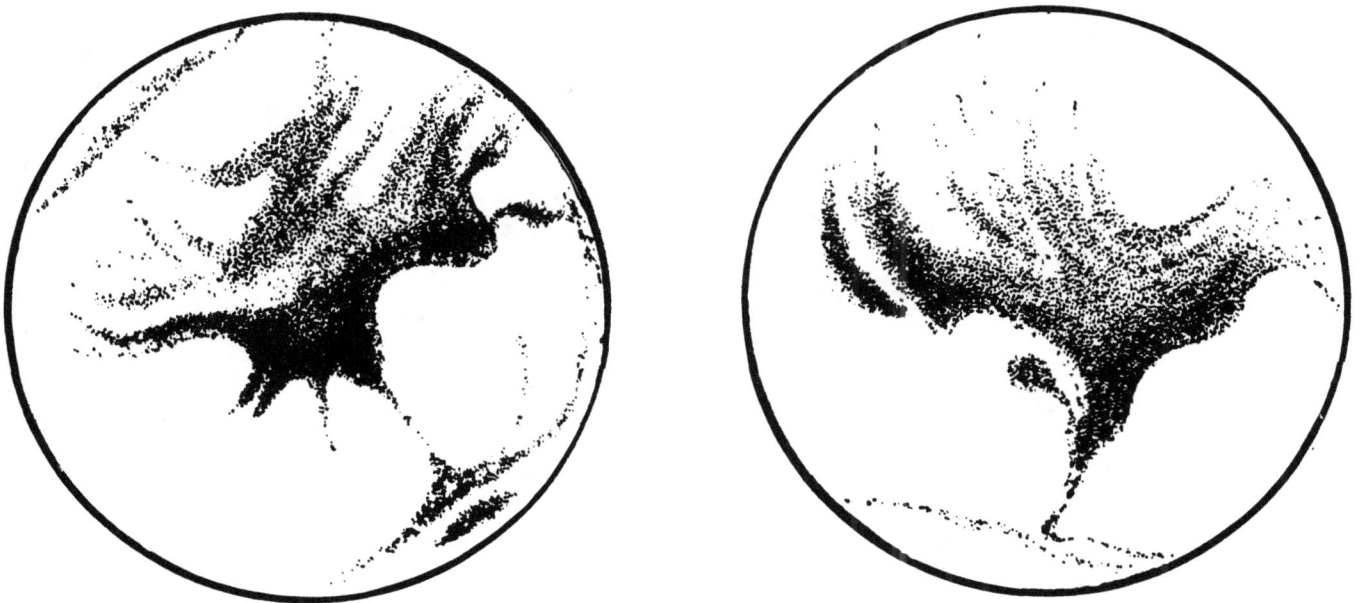

III *b*. Drawings of Mars made by the Reverend W. R. Dawes in 1864–1865. The drawing at left shows the region of Sinus Sabaeus, Margaritifer Sinus, and Aurorae Sinus; the one at right the region of Mare Cimmerium and Syrtis major.

43

IV *a*. Map of Mars by Richard A. Proctor, drawn in 1867. Following the usage generally adopted for the features of the moon, Proctor named the Martian features after astronomers, especially astronomers who had observed Mars.

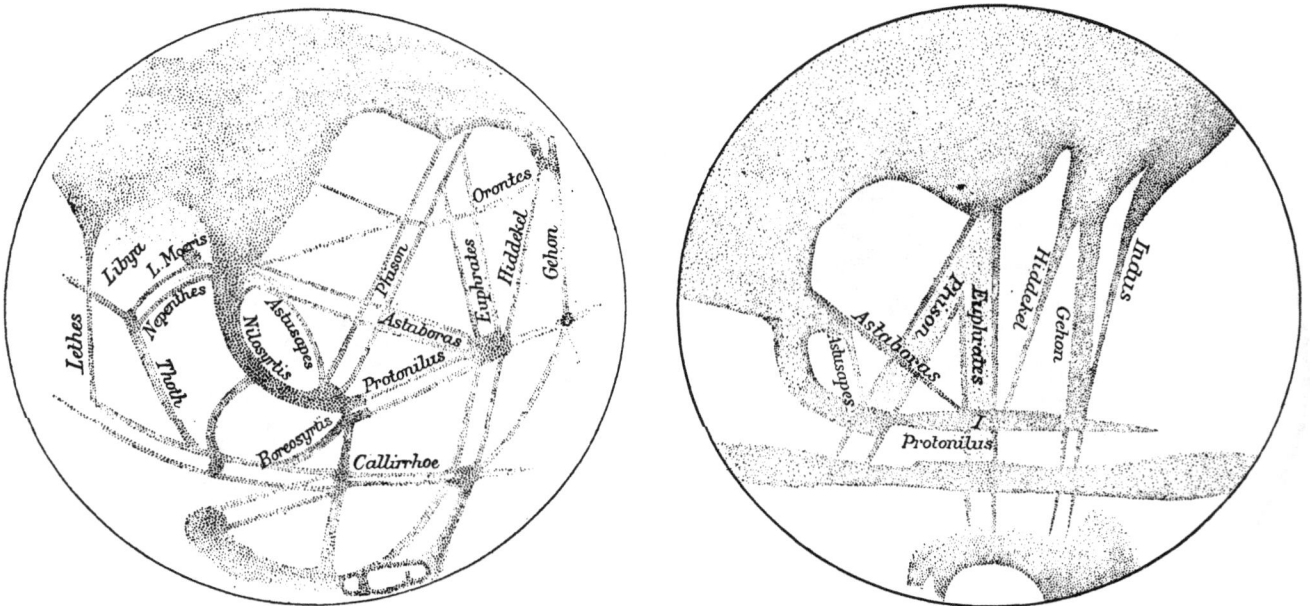

IV *b*. Two drawings of Mars made on June 4, 1888. The drawing at left was made in Milan by Giovanni Schiaparelli, the one at right at Nice one hour later by Henri Josef Perrotin. The difference in appearance is caused in part by the rotation of Mars during the one-hour interval, but also by the different styles of drawing of the two observers.

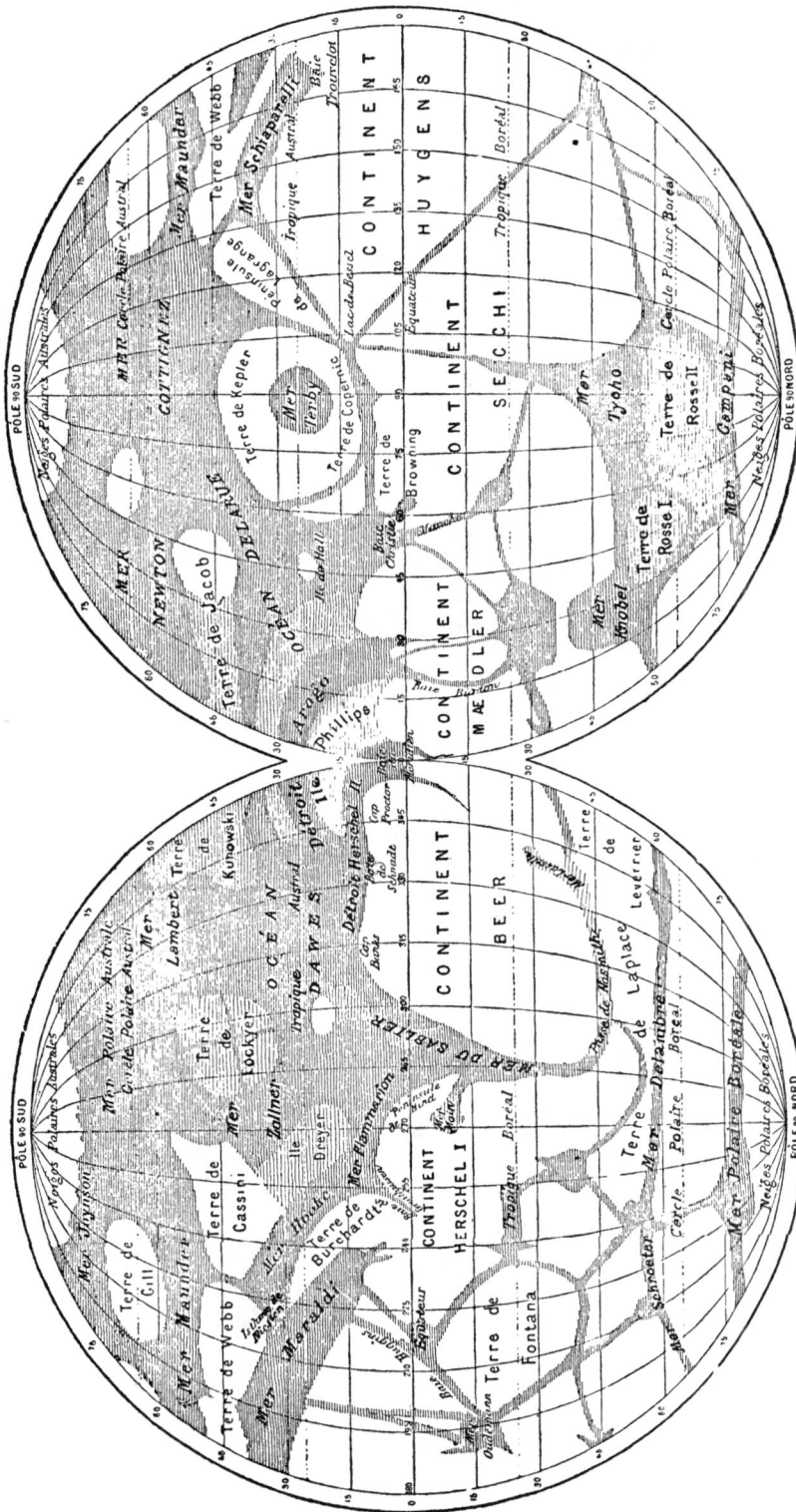

V. Map of Mars by Camille Flammarion, published in 1892. The language used on the map is, of course, French, but Flammarion followed Proctor's example in naming Martian features after famous astronomers, assigning the same name to the same area whenever possible.

45

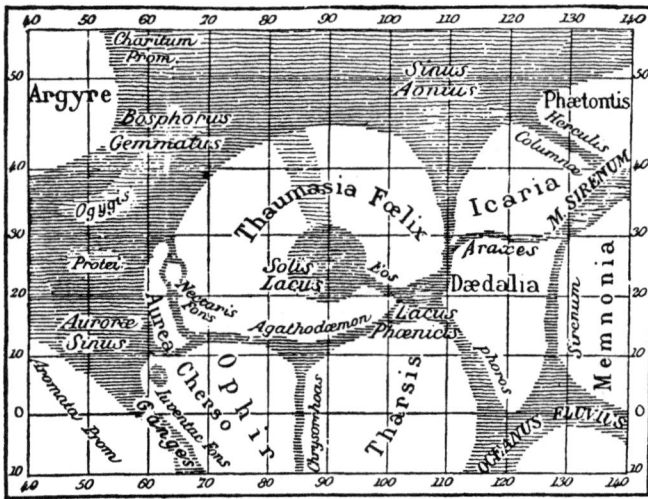

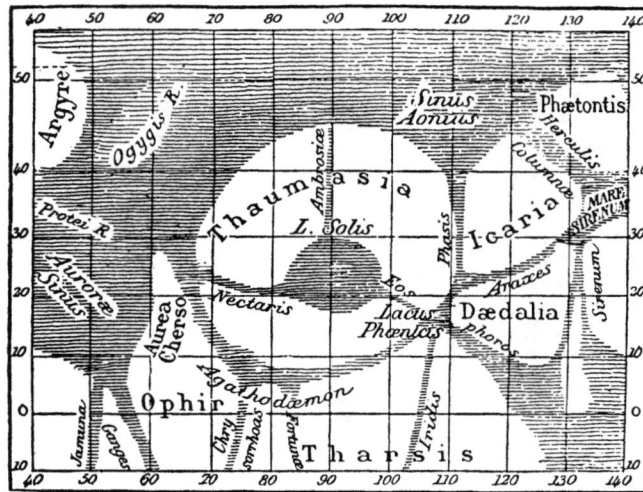

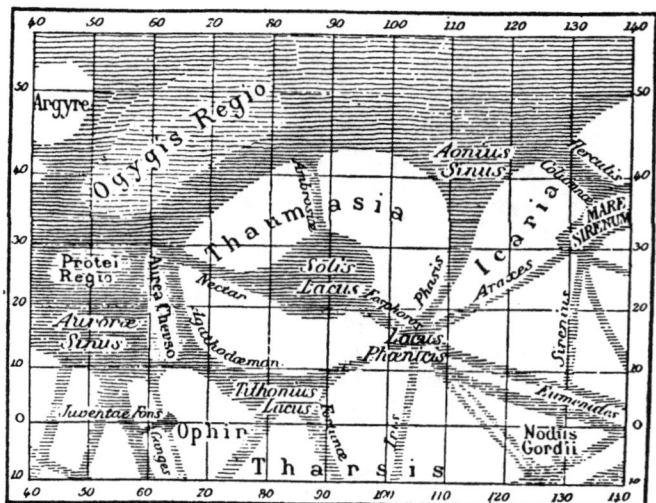

VI. The region of Lacus Solis: in 1877 (top); in 1879 (center); and in 1881 (bottom), according to Giovanni Schiaparelli. (From *La Planète Mars* by C. Flammarion.)

46

VII. Drawings of Mars made in February and March 1884 by E. L. Trouvelot at the Observatory of Meudon in France. These four drawings were selected by Trouvelot as the best of several dozen which he made during the opposition of January 31, 1884. They were first published in *L'Astronomie*, September 1884.

August 11, 9:10 P.M.

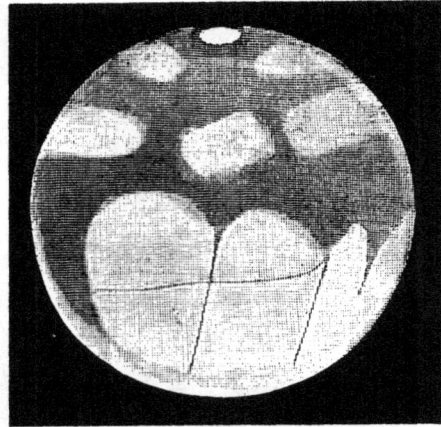

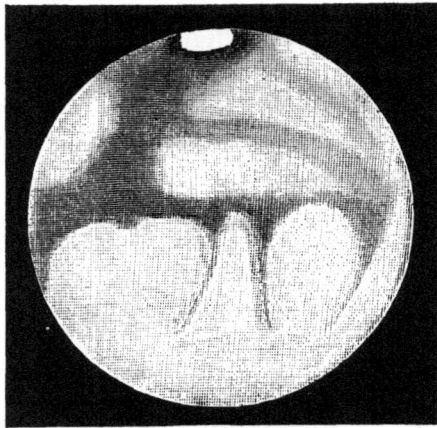

August 11, 10 P.M.

August 13, 11 P.M.

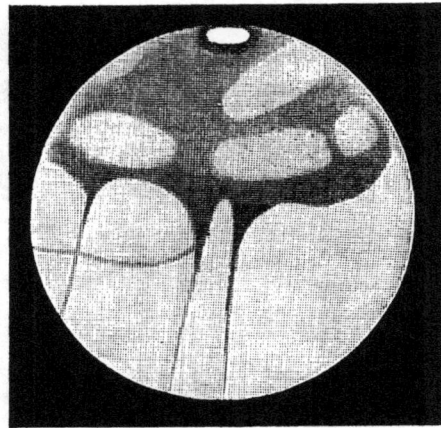

August 13, 11:45 P.M.

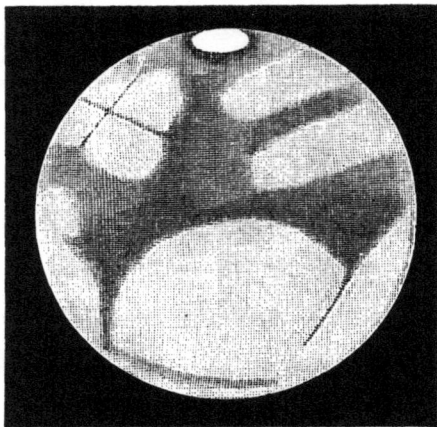

August 16, 10 P.M.

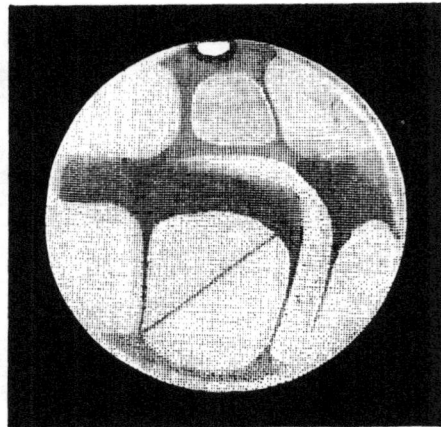

August 23, 9:15 P.M.

VIII. Drawings of Mars made during the perihelion opposition of 1892 at the Observatoire de Juvisy in France. The three drawings on the right and the bottom one on the left were made by Léon Guiot, the other two by M. Quénisset. (From *La Planète Mars* by C. Flammarion.)

While the one camp denied that the canali even existed, the other camp was perfectly willing to accept Schiaparelli's word for them. As regards the mistranslation of "canali" into "canals," they did not consider it a mistranslation at all but an explanation. Canals are artificial channels; their existence presupposes inhabitants intelligent enough to develop such works. This second camp was more diversified professionally than the first; in fact, astronomers were probably a minority in it, the others being representatives of other sciences who happened to be interested in astronomy.

Schiaparelli held himself fairly much aloof from the debates his announcement had caused. He was careful never to state outright that he considered the canali the work of intelligent beings, but he was equally "careful not to combat this supposition which includes nothing impossible." [3] After the opposition of 1879 Schiaparelli reported that something very curious had taken place: the canal he had called Nilus on his 1877 map had appeared double. There were two canals where only one had been before. This was the first instance of the phenomenon that later came to be called the "gemination" of the canals. But during that opposition Schiaparelli was mostly busy filling in on his map the regions to the north of the Martian equator which had not been positioned favorably in 1877. Though the planet was farther away during the next opposition of 1881, Schiaparelli had good results:

None of the dark lines called canali was missing. . . . Effects probably of solar nature laid bare an enormous amount of detail not even suspected during the previous oppositions. Wide regions which in 1879 appeared as indeterminate mists and seemed to belong to the so-called seas were resolved into most complicated intricacies of pure lines. Then the curious and unforeseen gemination of the canali was revealed.

The opposition of 1881 occurred late in December, the next one, of 1884, on January 31, the one following in 1886 on March 6, soon after Martian aphelion. It was at this opposition, otherwise not too favorable, that other observers reported "canals." Perrotin and Thollon, at Nice, drew a partial map of Mars showing more than two dozen different canals, about eight of these appearing double. The positions agree well with those of the canals on a Schiaparelli map; in fact, the map looks like a Schiaparelli map, but one drawn with a heavy hand. Other observers, in Belgium, in England, and in the United States, also saw canals during that opposition. All drawings agreed on one point: no "canal" ever petered out or stopped

3. *"To mi guardero bene dal combattere questa supposizione, la quale nulla include d'impossibile."*

short in a light area; they always went all the way from one dark area to another, and where several canali met they formed rather large round spots which Schiaparelli had called "lakes." Six years later the American astronomer William H. Pickering added the information that there was always a small spot where canals crossed each other; these were too small to be called "lakes," and Pickering suggested the term "oasis."

By 1892—a year with a near-perihelion opposition in early August—the concept of Mars had pretty well condensed into a composite picture. It was described by Schiaparelli in a long article which appeared in *Natura ed Arte*, February 15, 1893, and was translated into English by Pickering. Strange to say, this was the first direct translation of any of Schiaparelli's writings into English; everything that had been translated earlier had been first translated into French or German, and then into English. In this article Schiaparelli went over the Martian features one by one—or better, type by type—beginning with the white polar caps:

> . . . it is manifest that if the white polar spots of Mars represent snow and ice, they should continue to decrease in size with the approach of summer in those places, and increase during the winter. Now this very fact is observed in the most evident manner. In the second half of the year 1892 the southern polar cap was in full view; during that interval, and especially during the months of July and August, its rapid diminution from week to week was very evident, even to those observing with common telescopes. This snow (for we may well call it so) which in the beginning reached as far as latitude 70° and formed a cap of over 2000 kilometers [1240 miles] in diameter, progressively diminished, so that two or three months later little more of it remained than an area of perhaps 300 kilometers [185 miles] at the most, and still less was seen later in the last days of 1892. In these months the southern hemisphere of Mars had its summer; the summer solstice occurring upon October 13. Correspondingly the mass of snow surrounding the northern pole should have increased; but this fact was not observable, since that pole was situated in the hemisphere of Mars which was opposite to that facing the earth. The melting of the northern snow was seen in its turn in the years 1882, 1884 and 1886. . . .

About the Martian atmosphere Schiaparelli continued:

> In every climate, and under every zone, its atmosphere is nearly perpetually clear, and sufficiently transparent to permit one to recognize at any moment whatever, the contours of the seas and continents. . . . Here and there we see appear from time to time a few whitish spots changing their position and form, rarely extending over a very wide area. . . . It is possible that they may be layers of cloud, because the upper portions of terrestrial clouds, where they are illuminated by the Sun, appear white. But various observations lead us to think that we are dealing rather with a thin veil of fog, instead of a true nimbus cloud, carrying storms and rain. Indeed it may be merely a temporary

condensation of vapor, under the form of dew or hoar frost. Accordingly, as far as we may be permitted to argue from the observed facts, the climate of Mars must resemble that of a clear day upon a high mountain. By day a very strong solar radiation hardly mitigated at all by mist or vapor, by night a copious radiation from the soil towards celestial space, and because of that a very marked refrigeration. Hence a climate of extremes, and great changes of temperature from day to night, and from one season to another.

Finally Schiaparelli, after a fairly extensive description of his map which accompanied the article, progressed to the canali:

All the vast extent of the continents is furrowed upon every side by a network of numerous lines or fine stripes of a more or less pronounced dark color whose aspect is very variable. They traverse the planet for long distances in regular lines, that do not at all resemble the winding courses of our streams. Some of the shorter ones do not reach 500 kilometers (300 miles), others extend for many thousands, occupying a quarter or sometimes even a third of a circumference of the planet. Some of these are very easy to see, especially the one . . . designated by the name of Nilosyrtis. Others in turn are extremely difficult, and resemble the finest thread of spider's web drawn across the disk. They are subject also to great variations in their breadth, which may reach 200 or even 300 kilometers (120 to 180 miles) for the Nilosyrtis, whilst some are scarcely 30 kilometers (18 miles) broad. . . . These lines are the famous canals of Mars. . . . Their length and arrangement are constant, or vary only between very narrow limits. Each of them always begins and ends between the same regions. But their appearance and their degree of visibility vary greatly, for all of them, from one opposition to another, and even from one week to another . . . often one or more become indistinct, or even wholly invisible, whilst others in their vicinity increase to the point of becoming conspicuous even in telescopes of moderate power. [Our map] shows all those that have been seen in a long series of observations. This does not at all correspond to the appearance of Mars at any given period, because generally only a few are visible at once. . . . The canals may intersect among themselves at all possible angles, but by preference they converge towards the small spots to which we have given the names of lakes. For example, seven are seen to converge in Lacus Phoenicis, eight in Trivium Charontis, six in Lunae Lacus, and six in Ismenius Lacus.

As for the gemination, he reported, after stating that this happens very rapidly in a few days "or even perhaps only a few hours":

The two lines follow very nearly the original canal, and end in the place where it ended. One of these is often superposed as exactly upon the former line . . . but it also happens that both the lines may occupy opposite sides of the former canal, and be located upon entirely new ground. The distance between the two lines differs in different geminations, and varies from 600 kilometers (370 miles) and more, down to the smallest limit at which two lines may appear separated in large visual telescopes—less than an interval of 50 kilometers (30 miles).

Schiaparelli knew, of course, that his readers were especially eager to hear his opinion about the nature of the canals. Schiaparelli led into his answer by describing the melting of the snows around the Martian north pole:

We have already remarked that at the time of melting they appeared surrounded by a dark zone, forming a species of temporary sea. At that time the canals of the surrounding region become blacker and wider, increasing to the point of converting, at a certain time, all of the yellow region comprised between the edge of the snow and the parallel of 60° north latitude, into numerous islands of small extent. Such a state of things does not cease, until the snow, reduced to its minimum area, ceases to melt. Then the breadth of the canals diminishes, the temporary sea disappears, and the yellow region again returns to its former area. The different phases of these vast phenomena are renewed at each return of the seasons, and we have been able to observe them in all their particulars very easily during the oppositions of 1882, 1884 and 1886, when the planet presented its northern pole to terrestrial spectators. The most natural and the most simple interpretation is that to which we have referred, of a great inundation produced by the melting of the snows—it is entirely logical, and is sustained by evident analogy with terrestrial phenomena. We conclude therefore that the canals [here it would have been better to use "channels" in the translation] are such in fact, and not only in name. The network formed by these was probably determined in its origin in the geological state of the planet.... It is not necessary to suppose them the work of intelligent beings, and notwithstanding the almost geometrical appearance of all of their system, we are now inclined to believe them to be produced by the evolution of the planet, just as on the earth we have the English Channel and the Channel of Mozambique.

Although Schiaparelli had done more than any other man before him to change the map of Mars (and to make the planet intensely popular in doing so), his general concept of the planet, in 1893, was not too different from that of Herschel a century earlier. There were the large polar caps of ice and snow which melted completely each summer because the season lasted twice as long as on earth and the sun can climb about 1½ degrees higher over the poles. There was an accidental difference between the hemispheres. Near the southern pole was a large open sea, the Mare australe, which could accommodate the waters released by the melting of the ice cap without flooding. But near the northern pole there was no such large natural basin, so inundations took place every northern spring. In either case the presumably natural channels then carry the water to the equatorial regions, where it evaporates to condense at the pole that happens to have winter.

It was at about that time that an observer found a very typical canal located in one of the "seas." This could be explained in only one way: the "sea" could not be what its name indicated. An alternate explanation had already appeared in *Science* in 1888, five years before Schiaparelli wrote the article just quoted. It had been suggested there that the seas, since they did not look as one would expect a sea to look, were probably areas of vegetation. This explanation was strengthened by the fact that nobody had ever seen the reflection of the sun in any of the "seas"— something that would be bound to happen if they were really open water. Very soon an obvious corollary was added: the canals also were not open bodies of water a hundred miles and more in width, but lines of vegetation. The true canal, whether artificial or natural, it was said, was probably only a mile or even less in width and therefore invisible. But the belt of land irrigated by it, and bearing vegetation in contrast to the surrounding desert, could be seen from the earth. There was even a terrestrial example which could be used for comparison, one that, seen from Mars, might appear to be a very typical earth canal—namely, the Nile. Every year the Nile floods the area beyond its banks. Since this area is desert, the flooding is beneficial and things begin to grow. The consequences of the flooding of the Nile could be observed from Mars—if there were astronomically inclined inhabitants on Mars with telescopes of about the same power as our own—while the river itself would be invisible. This would have the additional result of straightening the image. Naturally the river does not flow in a straight line, but all the pockets formed by its meanderings would be "filled in" with vegetation and the whole would appear as a dark band of nearly uniform width and apparent straightness. All the facts observed and described by Schiaparelli, the darkening of yellow areas, the "filling up" of the canali, their gradual disappearance after the melting of the polar cap—the canal is then a barely visible trace hardly differing in color from the surrounding area, he had written—could be understood in terms of vegetation. Our own desert vegetation lies dormant when the desert is dry and its color is like that of the desert, differing only in shade. But as soon as moisture is available it springs to life quickly—"in a few days or even perhaps only a few hours"—becoming dormant again when the moisture is gone.

Schiaparelli himself thought more in terms of vegetation at a later date. In April 1910 a German scientific magazine published a long article by Dr. Svante Arrhenius of Stockholm, which tried to explain the Martian phenomena in an entirely different way (see Chapter 4). A Mr. Franz H. Babinger, knowing that

Schiaparelli read German well (he had studied under Johann Franz Encke, then director of the observatory in Berlin), sent a copy to Schiaparelli. He received a reply dated May 19, 1910, in which Schiaparelli wrote:

For my person I have not yet succeeded in formulating an organic whole of logical and credible thoughts about the phenomena of Mars, which are perhaps somewhat more complicated than Dr. Arrhenius believes. . . . I believe with him that the lines and bands on Mars (the name "canals" should be avoided) can be explained as the results of physical and chemical forces, always excepting certain periodical color changes which are likely to be the result of organic events of large magnitude, like the flowering of the steppes on earth and similar phenomena. I am also of the opinion that the regular and geometric lines (the existence of which is still denied by many persons) do not yet teach us anything about the existence of intelligent beings on this planet. But I think it worth-while if somebody collected everything . . . that can reasonably be said in favor of their existence.

Since Schiaparelli died on July 4, 1910, this letter [4] may very well be his last utterance on Mars.

Up to 1877 the planet Mars had not been really neglected, but it had not received much more attention than Venus, Jupiter, or Saturn, which, most of the time, are more conspicuous and also more accessible objects in the sky. The results of that particular opposition changed the focus of attention. Mars had become the most exciting and the most talked-about planet. And since astronomy as a whole had grown so much that specialization among astronomers had become a necessity, there emerged, along with moon specialists, solar-research specialists, and so forth, a group of Mars specialists. Among the first to become well known were Eugenios Michael Antoniadi in France and William H. Pickering and Percival Lowell in the United States.

4. Printed in *Kosmos*, 1910, p. 303.

4. OPINIONS, HYPOTHESES, AND THEORIES

MARS, as must now have become evident, is a difficult object. The planet is small, and rather far away even under the best of circumstances. On the average, a magnification of one hundred and fifty times makes it appear as large as the moon does to the unaided eye. This in itself would not be too serious a detriment since it is easy to increase magnification. But unfortunately we have to look through the earth's atmosphere, and that brings up the problem of "seeing." While the atmosphere above a high mountain top may be considered "clear," in the sense that virtually all the dust and smoke and most of the water vapor are below the observer, it is never "quiet." Or rather it is "quiet" at irregular intervals for a few seconds or at most some minutes at a time. These are the moments of perfect "seeing," and if there is a location on earth where these moments are frequent, it has still to be found.

Mars is a difficult object—and this is beautifully illustrated by the reaction of an interested layman who succeeds in having himself shown around an astronomical observatory and getting occasional glimpses through a medium-sized telescope. The interested layman may never have looked through anything bigger than a portable telescope—if that—but he knows what to expect, for he has several illustrated books on astronomy on his bookshelf at home. And for a while he is well satisfied. Venus, already low on the horizon after nightfall, shows the brilliant sickle shape he has expected to see. The mountains of the moon with the long dark shadows look just as they do on large photographs. The craters do have their central mountains and the Great Valley of the Lunar Alps shows up clearly. Reddish Jupiter hangs in the field of view of the telescope with pronounced cloud belts paralleling its equator, and of the four large moons three happen to be visible. Sat-

55

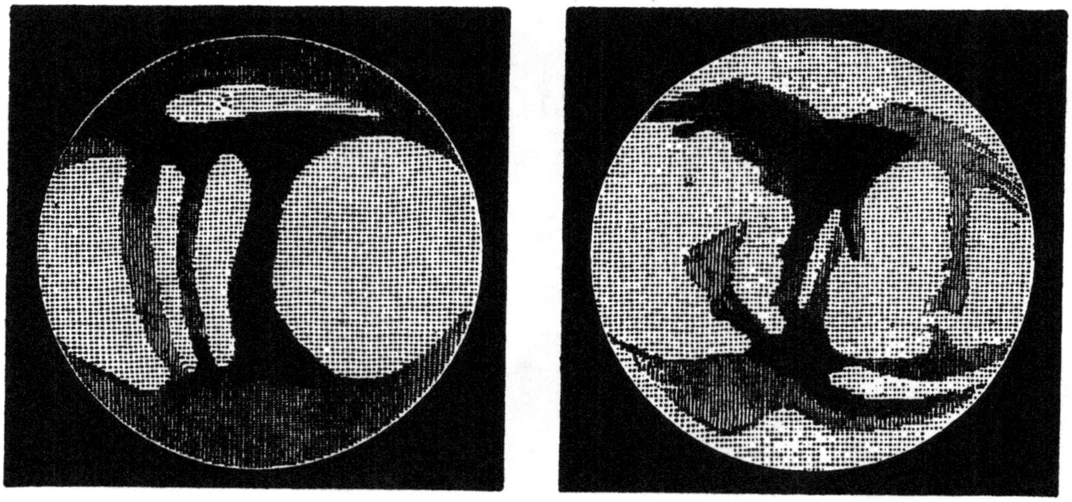

FIG. 12. A striking example of discrepancy even between skilled observers. The drawing at left was made at the Lick Observatory in the evening of July 27, 1888, by Edward Holden; the one at right by Keeler the same night, only a quarter-hour later, using the same instrument! A large-scale dust storm may explain most of the difference. (*After Flammarion.*)

urn's image with its wonderful rings is luminous and a thing of unearthly beauty against the dark background of the night sky.

But Mars—well, to put it plainly, Mars is a disappointment. The interested layman knows from his books that he cannot expect to see Schiaparelli's map in the sky. He has read the term "difficult object" on many occasions, but he nevertheless expected to see a little more. What can be seen in the telescope looks more or less like a brightly illuminated distant tangerine with a very white irregular spot on the top. The skin of the tangerine—to maintain this comparison for a short while longer—looks discolored: some sections are darkened with a greenish mold; others are a fairly light yellow. The first-time observer will realize after less than a minute why Mars photographs so poorly, and he is likely to reject a suggestion that he draw what he can see. The first impression of Mars through a medium-sized telescope on a good night is very much like the four drawings of Plate VII, though it should be remembered that a great deal of experienced skill is needed to make such drawings. While they correspond to first visual impressions, they certainly are not first impressions themselves.

No matter how confused and confusing the markings on the disk of Mars may look at first sight, the eye of the observer gradually learns to sort them out: Syrtis major is usually recognized first; then the area of Sinus Sabaeus and Margaritifer

56

Sinus becomes familiar. The "Lake of the Sun," Lacus Solis, is another conspicuous landmark, and usually Trivium Charontis shows clear and dark. All observers in all climates agree on these major features—on their shape and location, that is; not about their nature. But when it comes to the canali there is still disagreement. Some astronomers still say they don't exist and quote several well-known and admittedly capable observers as witnesses. Asaph Hall, for example, though he discovered the tiny and difficult satellites, never saw a canal. And France's great Mars specialist, E. M. Antoniadi, did not see any canals for many years; only late in life and with great reluctance did he admit that there were a small number of formations that might be called canals. Still another example of a planetary specialist who never saw a canal was Professor K. Graff in Germany. He made a special point of reiterating that he was not surprised at his "failure." Nothing reportedly as delicate as a canal could show "in that confusion of greyish, yellowish and brownish shades of endless detail near the limit of visibility." Professor Graff's attitude was that he would have seen canals if only they existed, and he implied that there was a little too much imagination on the part of people less rigidly self-disciplined than good conservative observers, such as, for example, Professor Graff.

The "canal men," on the other hand, in addition to Schiaparelli, were William H. Pickering and Percival Lowell. It is true that their pictures differed. Percival Lowell drew (and presumably saw) very fine lines, producing a most artificial-looking network. Professor Pickering drew soft and rather wide lines; only occasionally did he see fine lines, usually when a canal "geminated." But all the observers who have ever seen a canal agree that they are never there when you first take your place at the telescope. For a time, which may be quite long, Mars presents its customary picture, but then canals may "flash out," and three or four or five of them can be seen with the utmost clarity. They are there, clearly defined, "as sharp as an etching"—until the "seeing" turns poor again and they fade out.

The answer to the apparent riddle is probably just a matter of "seeing," of a temporarily well-behaved atmosphere along the line of sight. As has been said, a location where perfect "seeing" is common or even frequent has still to be found, assuming that one exists. But it is easy enough to think of a large number of places, and even of whole areas, where such "seeing" is unlikely to occur, and the geographical distribution of the "never-saw-a-canal" astronomers is at least suspicious. Professor Graff was the director of the Observatory of Bergedorf near Hamburg. Professor Asaph Hall worked in Washington, D. C., and E. M. Antoniadi was the director of the observatory at Meudon, France. But Giovanni Schiaparelli worked in Milan, Italy, William H. Pickering had his observatory in Jamaica, and Percival Lowell sat on a mountain top above the Arizona desert.

It speaks strongly for the reality of the canals that those that are seen, by different observers at different oppositions, always appear in the same places, the places where Schiaparelli first entered them on his maps. If they were pure imagination they would crop up anywhere. One explanation—the idea was first advanced by Vincenzo Cerulli and later adopted by Antoniadi—is that the canals are an "induced optical illusion," and that fine detail, arranged more or less along straight lines, cannot be seen separately and is perceived as a continuous line instead. This thought was checked experimentally by E. Walter Maunder, a Fellow of the Royal Astronomical Society, using some two hundred pupils of the Greenwich Hospital School. Mr. Maunder put emphasis on the fact that the boys selected, in addition to being keen-sighted, were "accustomed to do what they were told without asking questions; and they knew nothing whatsoever of astronomy, certainly nothing about Mars." (If these qualifications are needed, it would be awfully hard to repeat this experiment nowadays.)

Mr. Maunder continued: [1]

A diagram was hung up based upon some drawing or other of the planet made by Schiaparelli, Lowell or other Martian observer, but the canals were not inserted; only a few dots or irregular markings were put in here and there. And the boys were arranged at different distances from the diagram and told to draw exactly what they saw. Those nearest the diagram were able to detect the little irregular markings and represented them under their true forms. Those at the back of the room could not see anything of them and only represented the broadest features of the diagram, the continents and seas. Those in the middle of the room were too far off to define the minute markings, but were near enough for those markings to produce some impression upon them; and that impression always was that of a network of straight lines. . . .

Interesting as this is, it really doesn't prove much, even if the drawings came out exactly as described by Maunder. Not only is there an enormous difference between a keen-sighted schoolboy picked for obedience and ignorance and a mature observer with trained eye and hand, but the conditions must have been quite different. The experiment was presumably performed during school hours—that is, in daylight; the diagram was probably black and white, and the "dots and irregular

1. In his book *Are the Planets Inhabited?* London and New York, 1913, pp. 107 f.

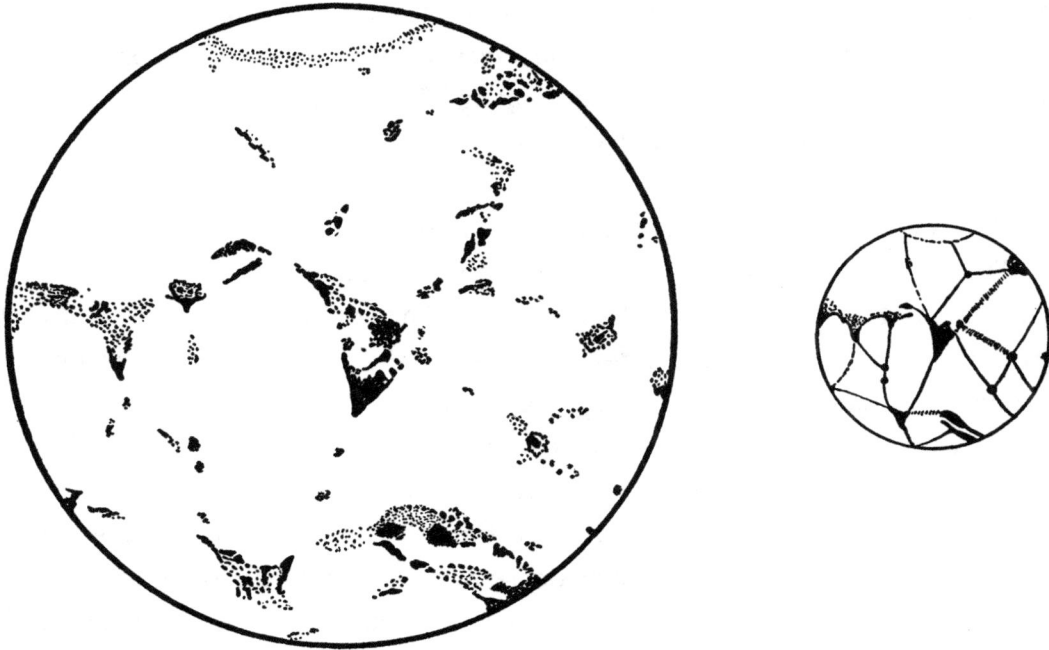

FIG. 13. One explanation for the "canals" of Mars is that they do not actually exist but are optical illusions caused by fine detail. The proponents of this explanation say that the planet appears to us like the small drawing at right, but *if* we could have an image as large as the drawing at left, we would see the detail composing the "canals," while the "canals" themselves would disappear.

markings" had been put in by the examiner. The similarity between this experiment and reality is at best the similarity existing between an artfully contrived crime story and a real crime.

Even Cerulli, who started this trend of thought, drew quite a number of canals himself, so that it is obvious that he saw them, in spite of his conviction that they were mere illusions caused by invisible detail. Supposing that he was completely right, it would still be an intriguing problem to determine what natural forces arranged fine detail approximately along straight lines. Moreover, the same effect should apply to other heavenly bodies too. One might, for example, look at the moon, using magnifications so low that the moon appears only as large in the telescope as Mars does under the highest magnifications practicable. The outcome should be a network of canals on the moon, for there is certainly no lack of detail on the lunar surface. Of course this has been tried, and some authors have published drawings intended to prove just this point. With the best will in the world these drawings cannot be said to be at all convincing. The fact remains that the canals, whatever their true nature may be, are exclusive with Mars. And as for

59

"imagination" and "wish fulfillment," the best answer is a frank statement by Dr. Robert S. Richardson of Mount Wilson Observatory: "If wishful thinking could conjure up canals, I'd have seen them long ago." But it took him until 1954 to see one with certainty, and probably two others.

Half a century ago quite a number of things about Mars could be considered as definite. The diameter was known and with it the area. By coincidence the total surface area of Mars happens to be just a little larger than the *land* area of earth. It was then known, too, that older ideas about the flattening of Mars near the Martian poles had been greatly exaggerated, and it was accepted that the difference between the equatorial and polar diameters of Mars amounted to about 20 miles. Recent work has put this figure a little higher again; Mars is somewhat more flattened than the earth. It was recognized then that the Martian atmosphere near the planet's surface was only about as dense as earth's atmosphere on top of a very high mountain; this has been revised downward rather drastically in the meantime. It was known that there is very little water on Mars, and it seemed likely that even the "Southern Sea," the Mare australe, was not a permanent body of water but at best a swamp. Since the known distance of the planet from the sun permitted the calculation of how much heat a pole could receive during spring and summer, it was also possible to calculate how thick a polar cap could be and still be melted completely by the heat available. The answer then given read: "Not more than five or six feet on the average, though in some places there might be much deeper drifts." This fifty-year-old statement, on the whole, is still more or less accepted, except that the figures have been revised downward drastically to read "five or six inches."

By the year 1900 the seasonal color changes of the darkish areas were also well established. In the course of each spring the darker areas deepened in shade and acquired a greenish tinge which disappeared again later in summer. With the melting of a polar cap, a "wave of darkness" spread equatorward, to fade out later, and to be repeated from the other pole as spring began there. Another well-established phenomenon was the "yellow clouds," which almost at once received the obvious explanation that dust from the yellow deserts was carried along by winds. One terrestrial feature was conspicuously absent: no mountains could be found. Most observers received the impression that the darker areas, the "seas," "lakes," "oases," and probably the canali too, were at a lower level than the yellow

60

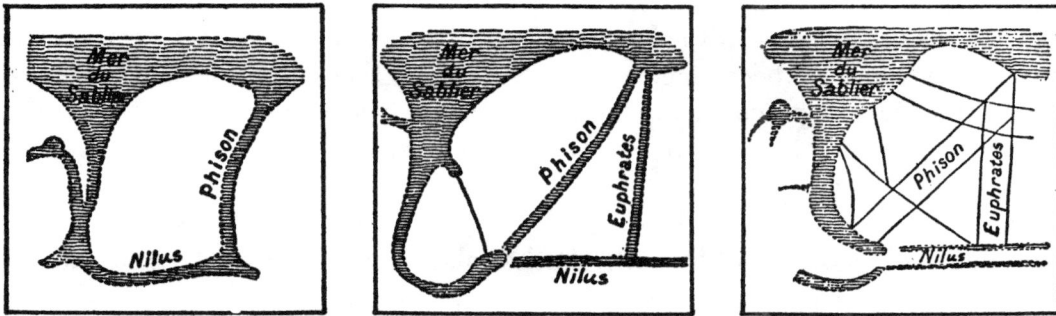

FIG. 14. Appearance of the region near the tip of Syrtis major during the oppositions of 1877 (left), 1879 (center), and 1882 (right). The canal labeled Nilus on these sketches appears as Protonilus on Schiaparelli's map. Mer du Sablier is the French name of Syrtis major. (*After Flammarion.*)

and reddish desert areas, but the deserts seemed flat. If there were mountains we should, at some time, see their shadows—unfortunately not when Mars is nearest us —and nothing like that could be found. Therefore, if any mountains are left on Mars they must be quite low, less than 1500 feet in height.

With all these major facts established, there remained the task of finding an explanation, of constructing an over-all picture of planetary conditions which accounted for all the observed facts, or at least contradicted none of them. The picture with the largest emotional appeal, and therefore the most popular, was the one defended by Percival Lowell in the United States and by Camille Flammarion in Europe.

It was the picture of a no longer hospitable world on which a supranational community of inhabitants struggled for survival by means of an irrigation system spanning the whole planet. The planet had grown unfriendly, but its thinking inhabitants—just because they could think and therefore invent and cooperate—were still alive. Percival Lowell, though not the original inventor of this idea and, as a matter of fact, slow to accept it, wrote about it with eloquence once he had convinced himself: [2]

Now, in the special case of Mars, we have before us the spectacle of a world relatively well on in years, a world much older than the earth. To so much about his age Mars bears evidence on his face. He shows unmistakable signs of being old. Advancing planetary years have left their mark legibly there. His continents are all smoothed down; his oceans have all dried up. *Teres atque rotundus,* he is a steady-going body now. If once he had a chaotic youth, it has long since passed away. Although called after the

2. In *Mars,* published in 1895.

most turbulent of the gods, he is at the present time, whatever he may have been once, one of the most peaceable of the heavenly host. His name is a sad misnomer; indeed the ancients seem to have been singularly unfortunate in their choice of planetary cognomens. With Mars so peaceful, Jupiter so young, and Venus bashfully draped in cloud, the planets' names accord but ill with their temperaments.

Although philosophers in the past, from Christiaan Huyghens to Immanuel Kant, had spoken of the inhabitants of other planets matter-of-factly and with calm assurance, Lowell knew that the skeptical nineteenth century would not accept the assertion of "people" on Mars without some raising of eyebrows. He therefore went over his reasoning:

We find, in the first place, that the broad physical features of the planet are not antagonistic to some form of life; secondly, that there is an apparent dearth of water upon the planet's surface, and therefore, if beings of sufficient intelligence inhabited it, they would have to resort to irrigation to support life; thirdly, that there turns out to be a network of markings covering the disk precisely counterparting what a system of irrigation would look like; and, lastly, that there is a set of spots placed where we should expect to find the lands thus artificially fertilized, and behaving as such constructed oases should. All this, of course, may be a set of coincidences, signifying nothing; but the probability points the other way.

Lowell then went on to say that the idea of intelligent inhabitants keeping alive by large-scale planning and engineering was merely the logical outcome of the higher age of the planet. Life on Mars, including intelligent life, would be not only relatively but also actually older than life on earth.

From the little we can see, such appears to be the case. . . . A mind of no mean order would seem to have presided over the system we see. . . . Party politics, at all events, have had no part in them; for the system is planet-wide. Quite possibly, such Martian folk are possessed of inventions of which we have not dreamed, and with them electrophones [telephones] and kinetoscopes [motion pictures] are things of a bygone past, preserved with veneration in museums as relics of the clumsy contrivances of the simple childhood of the race. Certainly what we see hints at the existence of beings who are in advance of, not behind us, in the journey of life.

The intelligent beings of Mars seemed to Lowell not only the best explanation of what he saw, but the only logical explanation—more logical, in any event, than some of the hypotheses advanced by others, to which Lowell gave short shrift, along with the objections against "people on Mars":

62

OPINIONS, HYPOTHESES, AND THEORIES

Startling as the outcome of these observations may appear at first, in truth there is nothing startling about it whatever. Such possibility had been quite on the cards ever since the existence of Mars itself was recognized by the Chaldean shepherds, or whoever the still more primeval astronomers may have been. Its strangeness is a purely subjective phenomenon, arising from the instinctive reluctance of man to admit the possibility of peers. . . . To be shy of anything resembling himself is part and parcel of man's own individuality. Like the savage who fears nothing so much as a strange man, like Crusoe who grows pale at the sight of footprints not his own, the civilized thinker instinctively turns from the thought of mind other than the one he himself knows. To admit into his conception of the cosmos other finite minds as factors has in it something of the weird. Any hypothesis to explain the facts, no matter how improbable or even palpably absurd it be, is better than this. Snow-caps of solid carbonic acid gas [carbon dioxide, the "dry ice" of current commerce], a planet cracked in a positively monomaniacal manner, meteors ploughing tracks across its surface with such mathematical precision that they must have been educated to the performance,[3] and so forth and so on, in hypotheses each more astounding than its predecessor, commend themselves to man, if only by such means he may escape the admission of anything approaching his kind. Surely all this is puerile, and should as speedily as possible be outgrown . . . [but] we must be just as careful not to run to the other extreme, and draw deductions of purely local outgrowth. To talk of Martian beings is not to mean Martian men. Just as the probabilities point to the one, so do they point away from the other. . . .

It is not easy to say in a sentence or two why most professional astronomers did not agree with Lowell and why his reasoning is now generally considered wrong in spite of the apparently logical development of his hypothesis. It was probably less a case of drawing wrong conclusions than one of operating with mistaken assumptions. In the first place, he assumed surface conditions still too much like those of earth. There is less air than he thought, but this fact alone would not spoil his argument. It is far more weighty that there is much less water than he thought. The arithmetic of so and so many square miles of vegetated area and so and so many more miles of canals versus so and so many square miles of polar cap did work out with a polar cap 6 feet thick on the average, but with a polar cap 6 inches thick it does not. Most important, the network of the canals seems to have appeared far more geometrical to him than it really is. Though some astronomers agreed with him all along and people outside the profession were, of course, utterly fascinated,

3. The ideas here ridiculed by Lowell were elaborated on by Johnstone Stoney and by Professor J. Joly of Dublin. The former asserted that the polar caps had to be solid carbon dioxide because there was no evidence of water, while the latter took the canals to be ridges caused by the gravitational attraction of asteroids skimming over the surface at a close distance.

63

a number of scientists tried their minds on the problem of devising a picture of Mars which agreed with the observed facts but did not require the assumption of intelligent inhabitants.

One of them was Professor Elihu Thompson, the famous researcher of the early days of electrical engineering. He built a telescope which was ready in time for the opposition of 1906; he not only saw the canals himself but had his family and his engineers make independent drawings which agreed well with one another and also with Lowell's sketches.

While Thompson did not believe in intelligent Martians, he still felt that the canals had been made by living beings. His reasoning, as reported by David O. Woodbury, who unfortunately failed to state his source, ran as follows:

> Though there is little water or oxygen on the planet, there may well be enough carbon dioxide to support considerable vegetation. There may even be primitive animal life. As the Martian seasons advance, the warm moist climate of the equator expands toward the poles, as it does on earth, taking the vegetation with it. This produces the annual migration of living things. There being no mountians or large rivers to obstruct them, the animals can make their yearly treks in straight lines, going to high latitudes on the little globe. Repeated fertilization and the long process of wear have gradually established paths of travel which have grown up with thick vegetation, leaving the rest of the planet a desert. At this great distance the pattern of interlacing routes can be mistaken for canals.[4]

In 1907 old Alfred Russel Wallace, who had conceived the idea of evolution independently of Darwin and by that very fact accelerated Darwin's publication of his own work, published a 110-page book which attacked Lowell rather sharply on a broad front. Wallace's book, according to his own words, had begun as a review of one of Lowell's books and had grown into a book itself. Wallace had not made any observations of his own and had to rely on published reports, mostly those of William H. Pickering, who was as devoted an observer of Mars as Lowell was. At one time Pickering had suggested that the canals might be natural cracks in the surface from which volcanic carbon dioxide and volcanic water vapor rose to support the vegetation we saw from earth. Wallace adopted this suggestion wholeheartedly and tried to show that cracks thousands of miles in length *must* form, if a still warm planetary crust cools off while resting on a planetary core which is

4. *Beloved Scientist—Elihu Thompson: A Guiding Spirit of the Electrical Age*, by David O. Woodbury (New York: Whittlesey House, 1944). Strangely enough, the German inventor Hermann Ganswindt had spoken of the canals as "routes of animal migrations" before Percival Lowell published his first book.

IX. Earth, Mars, and the moon, drawn to the same scale to show their relative sizes. Terrestrial desert areas are especially indicated for comparison with those of Mars.

X. The ruins of the Temple of Zeus Olympus near Athens, from the time of Hadrian. In such severely classical stone architecture a column 60 feet high had a diameter of 6 feet, and the clear span of the lintels bridging a colonnade was 9 feet . . .

. . . but stone architecture on Mars would look spidery to terrestrial eyes. The diameter of a 60-foot column would be 2 feet and the clear span of the lintels would be 27 feet. However, if the same type of stone were used, such a structure would be just as "massive" under the weaker Martian gravity.

XI. One theory about the canals of Mars is that they are long low valleys, natural faults in the Martian crust through which fogs roll from the polar areas when spring comes to one or the other pole.

XII. Mars as it appears to the naked eye when seen from its outer moon Deimos.

XIII. Mars as it appears to the naked eye when seen from its inner moon Phobos.

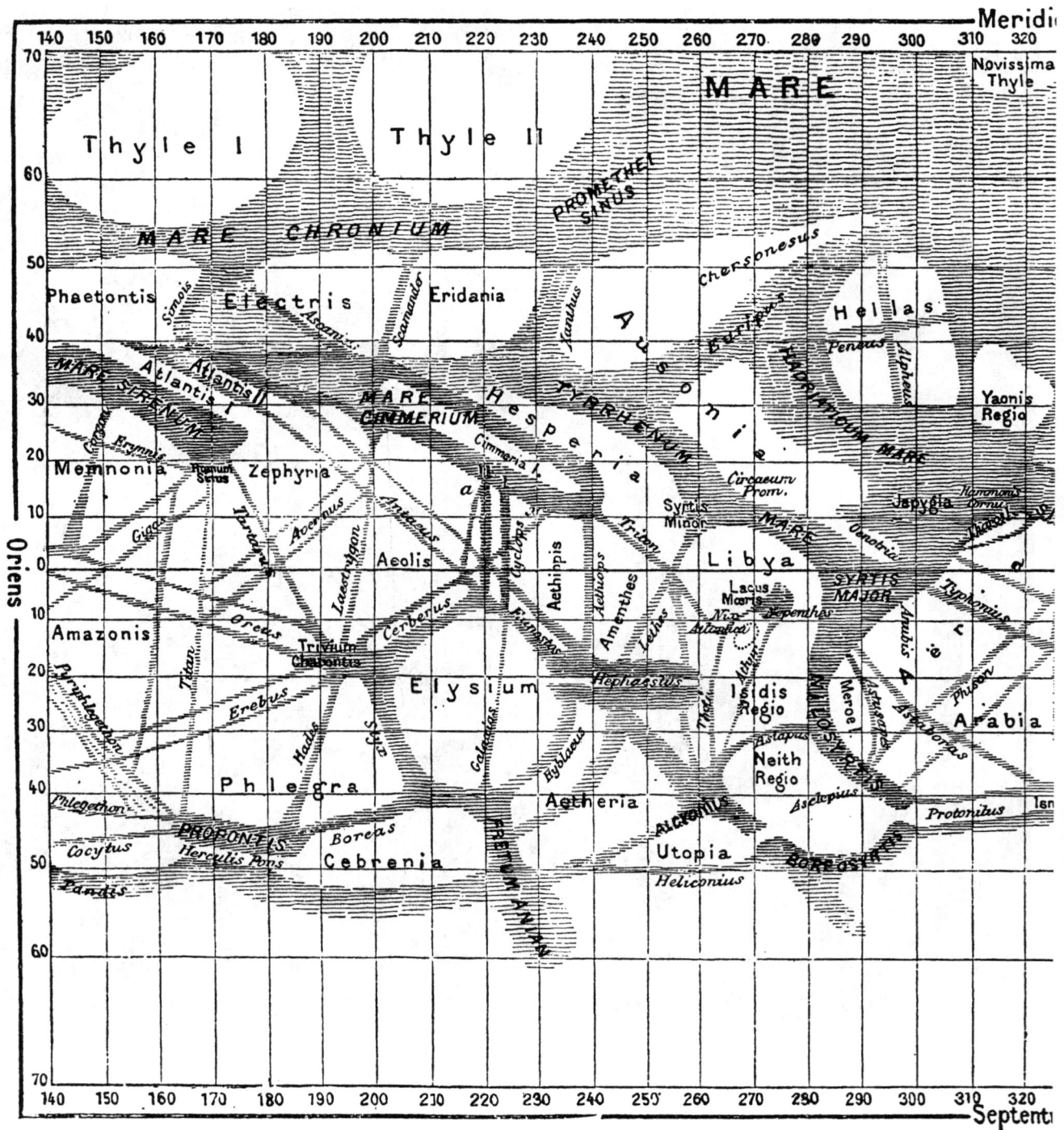

XIV-XV. Giovanni Schiaparelli's map of Mars, compiled over the period from 1877 to 1886. Schiaparelli's names were mostly based on classical geography, or were simply descriptive terms, for example, Mare australe (Southern Sea). Most of Schiaparelli's names are still in use.

70

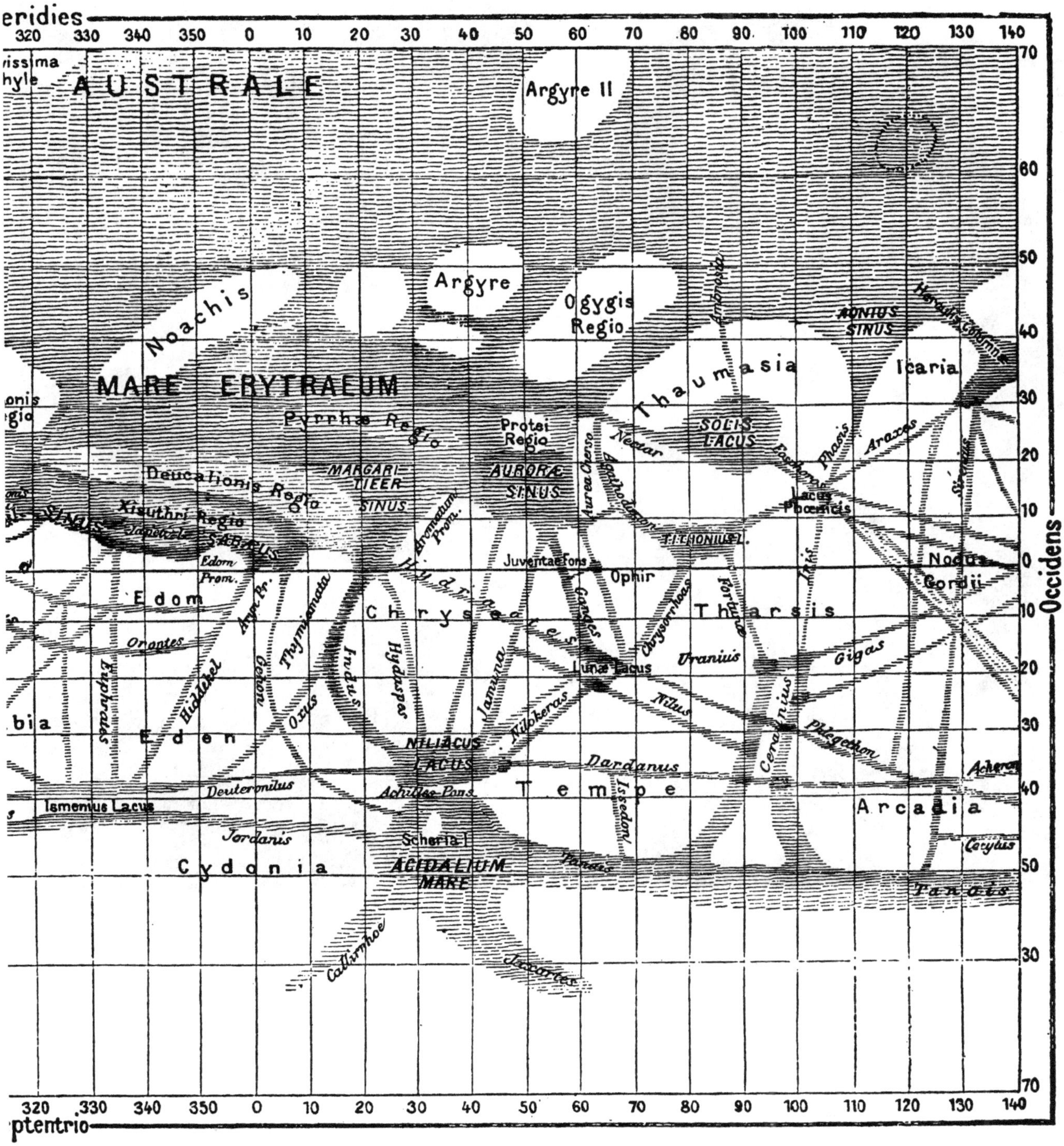

eridies
AUSTRALE
Argyre II
Noachis
Argyre
Ogygis Regio
Ambrosia
AONIUS SINUS
Herculis Columnae
Icaria
MARE ERYTRAEUM
Thaumasia
Pyrrhae Regio
Protei Regio
SOLIS LACUS
Nectar
Araxes
Phasis
Deucalionis Regio
MARGARITIFER SINUS
AURORAE SINUS
Aurea Cherso
Agathodaemon
Lacus Phoenicis
Sirenius
Xisuthri Regio
Aromatum Prom.
TITHONIUS L.
Juventaefons
Nodus Gordii
Edom Prom.
Ophir
Edom
Chrysorrhoas
Tharsis
Occidens
Chryse
Ganges
Fortunae
Oronte
Hiddekel
Gehon
Thymiamata
Indus
Hydaspes
Jamuna
Lunae Lacus
Uranius
Gigas
Euphrates
Oxus
Nilokeras
Nilus
Ceraunius
bia
Eden
NILIACUS LACUS
Phlegethon
Dardanus
Deuteronilus
Achilles Pons
Tempe
Issedon
Ismenius Lacus
Arcadia
Jordanis
Scheria I
Cydonia
ACIDALIUM MARE
Tanais
Callirrhoe
Jaxartes
ptentrio

XVI. A Martian landscape in the vicinity of the southern pole.

already cold and no longer contracting. Then, of course, he had to prove that Mars did have a noncontracting core inside a still-contracting rind. Since this led into problems of planetary origin, the argument occasionally got somewhat out of hand.

Furthermore, Wallace continued, even if this explanation is not accepted, Lowell is still wrong on every count. In the first place, "all physicists are agreed that, owing to the distance of Mars from the sun, it would have a mean temperature of about *minus* 35° Fahrenheit, even if it had an atmosphere as dense as ours." In the second place, it is clear from the observations that Mars does not, in reality, have a temperature as high as the freezing point of water. In the third place, water vapor cannot exist in the atmosphere of Mars; hence there can be no water at all on Mars. The book concluded: "Mars, therefore, is not only uninhabited by intelligent beings such as Mr. Lowell postulates, but is absolutely UNINHABITABLE."

The book which followed in chronological order after Wallace's blast was, in many respects, even more surprising. It appeared in Zürich in 1909 and bore the title: *Explanation of the Surface of the Planet Mars.* Its author was one Adrian Baumann, who almost literally turned all arguments around and the whole picture of Mars upside down. Aside from Wallace, who would not permit any water on Mars, everybody had been in reasonable agreement that there was very little water on Mars. Wrong, said Baumann, Mars has at least as much water as has earth, only almost all of it stays frozen all the time. In fact, the areas called deserts by everybody are "deserts" only in the semi-symbolic extended meaning of the word, for they are frozen oceans. The darkish areas are the land areas and are possibly vegetation; they might even harbor animal life. Islands in the frozen oceans are largely volcanic and active; when they do erupt we see the "yellow clouds," and it is the fall-out from these clouds which colors the frozen seas yellow and reddish. The white areas that have been seen fairly often, and the polar caps, might be considered hoarfrost or clouds of ice crystals. The canals that run from the still-active volcanic islands to the true land, the dark areas, are wide cracks in the frozen seas, which are only logical as the outcome of volcanism in a large mass of solid ice.[5]

5. The idea that Mars is a watery world had been expressed once before, in a book by Ludwig Kann, published in Heidelberg in 1901. Its title was *New Theory about the Origin of Coal and Solution of the Mars Problem.* The author probably read somewhere that some paleontologists believed that at least some coal beds were formed by large accumulations of aquatic plants, as exemplified in our time by the Sargasso Sea. Taking this as an established fact, he reasoned as follows: Mars is completely covered with water and most of the shoreless ocean is covered by floating masses of yellowish-reddish weeds of about the same type as our sargassum weed. The so-called "seas," the dark areas, are areas where the weed fails to grow for unknown reasons so that we see the dark bottom showing through fairly clear water. The canals are due to ocean currents which periodically part the masses of floating weeds!

THE EXPLORATION OF MARS

The next man after Baumann who tried to explain the features of Mars was again a famous scientist, at least as famous as Alfred Russel Wallace, though in an entirely different field. He was Svante August Arrhenius, physicist and chemist, winner of the Nobel Prize in chemistry in 1903 and, beginning in 1905, director of the Nobel Institute for Physical Chemistry. His opinion about our neighbor in space —"Mars is indubitably a dead world"—with the reasoning behind this opinion was published in various places, for the first time in the German *Kosmos*, 1910, pages 123-28. This was the article that the aged Schiaparelli read and commented upon shortly before his death.

In one respect Arrhenius agreed with both Wallace and Baumann: the temperature on Mars must be so low that water is normally frozen. He wrote:

As long as 20 years ago Christiansen in Copenhagen calculated that the sun's rays could not keep the Martian temperature above minus 37° centigrade [minus 34° F.]. The same calculation leads to a mean temperature of 6.5° centigrade [44° F.] for earth which is about 9° centigrade [16.5° F.] below the true value. One could hope that the temperatures on Mars are actually somewhat above the calculated values but that the difference should amount to 30° or 40° centigrade [up to 72° F.] could hardly be assumed.

Arrhenius admitted that locally at noontime the temperature may rise above freezing point, which, in conjunction with the low atmospheric pressure, could cause rapid evaporation or melting of snow.

The deserts of Mars, Arrhenius took to be very old geologically. Because of their great age they had had time to accumulate considerable quantities of cosmic dust, mostly iron, which oxidized and thereby caused the typical red color of the planet. The so-called lakes were just low areas and the canals earthquake cracks. But they did not become visible because of vegetation; the whole process of color change, the famous "wave of darkness," was a purely chemical phenomenon. He assumed that in low-lying areas there might be occasional true lakes, "like the desert lakes on earth very shallow, with very salty water and often evaporating completely." When such a lake does evaporate, the least soluble salts, the sulfur salts, will appear first at the shores in crystal form. Then ordinary salt and magnesium chloride will follow suit. In the center there will still be liquid water, or rather a saturated watery solution of calcium chloride, which does not freeze until the temperature has dropped below minus 65° Fahrenheit. But finally this solution freezes too, and the ice crystals evaporate because of the general dryness of the atmosphere and are carried to the coldest part of the planet, the pole which happens to be having winter, where they form a polar cap. When spring comes and the polar cap evaporates, the water is attracted by the usually highly hygroscopic salts which then appear dark again. And, concluded Arrhenius, if one makes the assumption that some water vapor, mixed with carbon dioxide, sulfur dioxide, and hydrochloric acid, still comes from the interior in some places, the wave of darkness

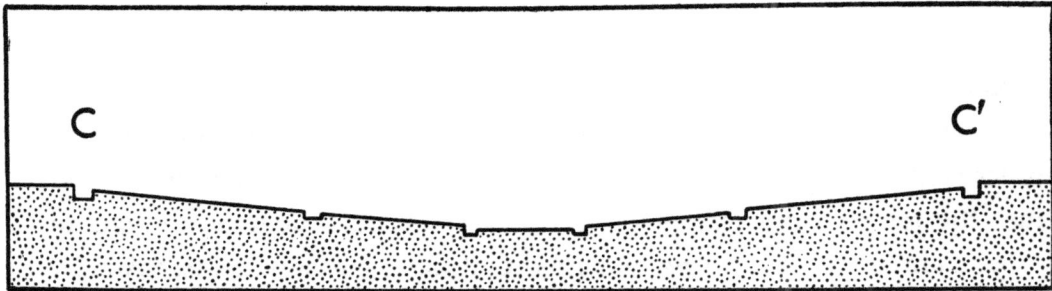

FIG. 15. Although Schiaparelli never stated definitely that he considered the "canals" of Mars artificial, he seems to have toyed with the idea. This is a redrawn version of a sketch he published in 1895 as a possible explanation for the doubling of the canals. The idea was that first the two outermost canals (C and C') would be filled until they overflowed to fill the next two canals downslope, and so forth to the center. As long as only C and C' carried water, the canal would appear doubled.

can be explained by two partly interlinked causes. One is the direct darkening action of moisture, the other a chemical reaction which changes the reddish iron oxides into black sulfides.

Mars, according to Arrhenius, then, was mostly cold desert, and where it was not desert it consisted of cold salt marshes, without any life but with seasonal color changes due to chemical trickery. As could be expected of the director of the Nobel Institute for Physical Chemistry, the professional reasoning was superb, but whether the explanation was the true one depended on a number of factors that still had to be established. The most important of these was the actual surface temperature on Mars. It could be calculated, of course, but every method of calculating it involved making a few assumptions about which one could not be sure. Moreover, nobody could swear that he had really taken *all* the factors into account; there might be some which would be easy to understand once they were pointed out but which could not be guessed in advance. In short: it was necessary to measure the surface temperatures of Mars instead of guessing about them.

THE EXPLORATION OF MARS

After the very fine opposition of 1924 this recurrent question of the temperatures on Mars at long last received an answer. Scientific Paper No. 512 of the Bureau of Standards, finished on April 28, 1925, reported on radiometric measurements made at the Lowell Observatory, Flagstaff, Arizona, during 24 nights in the months of July to September, 1924. Both the temperatures of the bright yellow areas near the center of the disk and those of dark areas like Syrtis major had been measured, at sunrise, at sunset, and at noon Martian time. The figures were a surprise to many theorists and presumably a relief to some; they turned out to be much higher than expected in some quarters.

At sunrise at opposition the temperature of a given area measured minus 45 degrees centigrade or about minus 49 degrees Fahrenheit. At the sunset side of the planet the temperature was precisely freezing point, 0 degrees centigrade or 32 degrees Fahrenheit. It must be remarked that these measurements, taken from the outside through the Martian atmosphere, could not be completely accurate. An area that measured freezing point through the Martian atmosphere would actually have a somewhat higher temperature.

At the center of the disk—Martian noon for that spot—there was a considerable difference between light desert, presumably high ground, and dark lowland. The desert temperatures at noon ran from minus 10 degrees centigrade [14 degrees Fahrenheit] to +5 degrees centigrade, or 41 degrees Fahrenheit. In the dark areas the noon temperatures ran from +10 to +20 degrees centigrade, or from 50 to 68 degrees Fahrenheit. In other words, for a human being the noon temperature in a dark area would actually be pleasant. But only the noon temperature. The average for the whole disk during the day was minus 30 degrees centigrade, or minus 22 degrees Fahrenheit. The nighttime temperatures could not, of course, be measured but they could be estimated from the early-morning temperatures: the average for the whole night side must be minus 70 degrees centigrade, or minus 94 degrees Fahrenheit. W. W. Coblentz, who wrote the paper, added:

The observed high surface temperatures may be accounted for on the assumption that the dark areas contain vegetation having the properties of the tuft-forming grasses of our high prairies and the tussock mosses and lichens of our dry tundras, which have a high absortivity for solar radiation and a low thermal conductivity. . . . The radiometric observations indicate that during the summer season on Mars temperature conditions at noonday are not unlike the bright cool days on this earth, with temperatures ranging from 5-15° centigrade or 40 to 60° Fahrenheit.

OPINIONS, HYPOTHESES, AND THEORIES

Even in astronomy there are what for want of a better term may be called "fashions." The very fine opposition of 1924 had been well observed, and by the time all the material gathered then was published the fairly good opposition of 1926 came around. But after its material was published the literature on Mars grew visibly scarcer. In part this was due to the cycle of oppositions, but other astronomical interests contributed heavily. Mars, which had held the center of the astronomical stage with only minor interruptions since 1877, was eclipsed by distant stars which had suddenly acquired great significance. A typical astronomical symposium of the 1930s talked about Cepheid variables, Wolf-Rayet stars, extragalactic nebulae, atomic transformations in stars, and everything else except the planets of our solar system. It is very typical that when Robert S. Richardson questioned fellow astronomers about their plans for a forthcoming opposition at the time of the Second World War, an astonishingly large percentage of them were unaware of the fact that an opposition was coming up.

After the war this attitude changed again, and since several fine oppositions were in the offing, a Mars Committee was organized to coordinate research work and collect research results. The committee met for the first time in October 1953 at the Lowell Observatory. It was functioning for the opposition of 1954 and will be quite active for that of 1956.

Since there is no new over-all theory about Mars to be reported, the best way of presenting what is now known and believed is feature by feature.

First, the Martian atmosphere: it is thin; the best determinations permit the conclusion that the pressure at the surface is somewhere between 62 and 70 millimeters of mercury, which corresponds to that at a height of between 16.5 and 17.5 kilometers (10.3 and 11 miles) in earth's atmosphere. At these heights our air has the prevailing temperature of the lower stratosphere—minus 55 degrees centigrade, which corresponds to minus 67 degrees Fahrenheit. At the Martian surface, just because it is at the surface, the temperatures are different, about the figures given by Coblentz in 1925. The problem of the composition of the Martian atmosphere is even more difficult than that of its pressure. The spectroscope, which is the instrument that could answer such a question, is handicapped by the fact that the light rays also have to traverse the terrestrial atmosphere, which, of course, produces much heavier lines. One of the most recent estimates is that the Martian atmosphere is about 96 per cent nitrogen and 4 per cent argon. The presence of carbon dioxide has been shown by the spectroscope; in fact, the percentage of carbon dioxide in the Martian atmosphere is higher than in the terrestrial atmosphere. Traces of water vapor seem to occur occasionally, but most of the time the water in the Martian atmosphere takes the form of floating ice crystals which the spectroscope cannot detect. Whether there is any oxygen present is not established.

It is almost traditional by now to say that the oxygen that was formerly present in the Martian atmosphere has been tied up chemically in the planet's crust. Rupert Wildt of Princeton some time ago made an interesting suggestion as to how this may have come about. Higher up in earth's atmosphere the ultraviolet radiation coming from the sun changes a percentage of the oxygen present into the tri-atomic form of oxygen known as ozone. Because of the lower pressure in the Martian atmosphere, such an ozone layer would have been near the ground, and since ozone is far more active chemically than normal oxygen is, it would have been used up in oxydizing the surface about as rapidly as it was formed. The oxygen, therefore, was chemically tied up, with the detour of ozone formation, but in this case the detour is the far more rapid route.

Physically the most extended feature of Mars is, of course, the desert which covers three-quarters of the surface. Except for Baumann and a few other theorizers who followed more or less in his footsteps, there has never been any serious disagreement about its nature, namely that it is desert colored by a fairly high percentage of iron compounds, a large part of which may even have come from space, as Arrhenius thought.

In regard to the polar caps there is agreement too, namely that they are frozen water, but the old question "how much frozen water" is still with us. Modern estimates run literally from 1/10 inch to 10 inches; to say "a few inches" is still a safe and probably correct statement. Whether this is snow or ice is a different question. *Most astronomers thought of rather fluffy snow, because the low gravity of Mars* offers little inducement for compacting. But the Russian astronomer G. A. Tikhoff decided after a study of pictures obtained during the opposition of 1939 that the poles were likely to be ice covered by hoarfrost. His compatriot V. V. Sharonov agreed with him, and there is little evidence on which to base a contradiction. One aspect which was fought over for a long time seems to have been definitely settled by Gérard de Vaucouleurs, the current Mars specialist among French astronomers. Many observers had seen a "dark fringe" around the melting ice cap, and most of

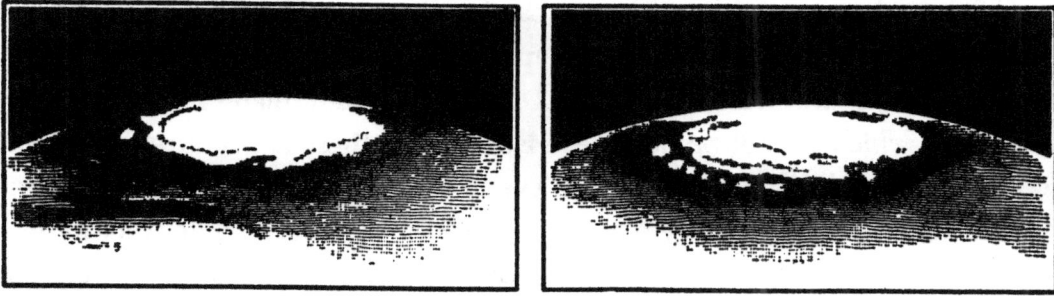

FIG. 16. The southern polar cap of Mars, as drawn in 1877 by Nathaniel E. Green at his private observatory on Madeira. The picture at left shows the appearance on September 1, the one at right that on September 8.

them took it to be very shallow water or else moist ground. But then it was proclaimed that the perspective was all wrong, that the dark fringe did not have the shape it should have, and that, in all probability, it was merely an optical contrast to the blinding white of the ice cap; and anyway in the low atmospheric pressure the ice cap would evaporate rather than melt. After seeing the dark fringe clearly and repeatedly, de Vaucouleurs came to the conclusion that it actually is moistened soil and that the ice really melts, *after* a considerable amount has evaporated and therefore the air over the polar cap is temporarily not completely dry. This explanation fits in well with the fact that the only spectroscopic observations which did seem to indicate the presence of water vapor with any clarity were taken over the melting ice caps.

The next feature is the clouds in the Martian atmosphere. There are either two or three types—this still remains to be settled—plus a mysterious something which is called "layer" or "mist" in order to give it a name at all. The biggest and most common and also most easily explained phenomenon is the "yellow clouds" over the deserts, which are what their name and the circumstances suggest: clouds of fine dust carried before a wind. The other cloud phenomenon is known as "blue" clouds or "blue-white" clouds, and the problem is whether or not these are two different types. Bright white clouds were often reported by the visual observers of the last century, and when they showed up, brightly illuminated by the sun, near the rim of the disk, they were even sometimes dubbed signals of the Martians.

The name "blue" clouds should not be taken literally; the color designation refers to the color filter used in photographic work. When Mars is photographed through color filters, pictures taken through yellow or orange filters look very much like the visual impression. A picture taken through a red filter also resembles the visual impression, except that the polar cap hardly shows. But through blue or violet filters most markings simply vanish (except for the polar cap, which shows strongly), and the planet as a whole seems larger. On such blue filter plates faint clouds sometimes show which cannot be detected visually, and these are referred to as "blue" clouds. But if a cloud is visible to the eye in the same spot it appears as a bright white cloud. Naturally a really white cloud would show on a "blue" picture just as does the polar cap; the question is whether the "blue" clouds are merely white clouds too faint to be seen directly, or whether they are a different kind. The white clouds visible to the eye are generally agreed to consist of ice crystals.

The explanation for the fact that the image of Mars looks slightly larger through a blue filter is that red and yellow light penetrates the Martian atmosphere all the way to the ground (unless temporarily blocked by a dust storm), while the shorter wave lengths of blue and violet light do not. A blue-filter picture, then, is a picture of the atmosphere rather than of the planet, or, more precisely, a picture of the scattering of blue light in the atmosphere. But with the atmosphere of Mars what it is, so much scattering simply should not take place. One therefore has to assume an atmospheric layer which has "an astonishing power of diffusion and absorption of short waves," to use E. C. Slipher's words. This is the "blue mist" or the "violet layer." It has been suggested that this consists of very fine ice crystals, but this is conjecture pure and simple. That it must be a special layer which does the fantastic scattering is proved by the fact that every once in a while there is a "blue clearing." When this occurs, surface features can be photographed even through a blue filter; a blue-filter picture of Mars then looks like a yellow-filter picture. A most interesting point is that a blue clearing has always occurred at opposition—at least for as long as astronomers have been watching for one. Since at opposition the earth is between the sun and Mars, E. P. Martz has suggested that the clearing may actually be caused by the earth, or rather by the earth's magnetic field, which might alter the solar radiation reaching Mars. If so, a "blue clearing" would be an anomaly for Mars, which would be temporarily deprived of its normal, if mysterious, "blue mist."

One specific cloud deserves separate mention, even though the case is not as clear cut as one should wish. On January 15, 1950, the Japanese astronomer Tsuneo Saheki observed a large dark gray-yellowish cloud in the area of Eridania and Electris, forming a circular patch some 450 miles in diameter and rising to an estimated height of between 60 and 100 miles. American and European observatories could not observe Mars at that time, and when night came for them the cloud had apparently disappeared. But other Japanese observers confirmed the report. On March 29 of the same year another Japanese astronomer, S. Ebisawa, saw a similar gray cloud over the southeastern part of the Mare Sirenum. It could still be seen on April 2, but had changed color and looked dull bluish-white. A third such cloud was seen over Eridania in February and early March 1952 and a fourth in the same general location on April 16, 1952. It was first noticed by Ebisawa and confirmed by Saheki. These clouds differed from the "normal" yellow clouds in two respects: their color was different and they rose to a great height. A single such cloud could easily be explained as the telltale sign of a crash of a very large meteorite or small asteroid, but since all four of them occurred in an area not much more than 500 miles in diameter it is more likely that they were of volcanic origin. It may be mentioned that Antoniadi, in 1909 and 1911, repeatedly saw clouds which he thought to be volcanic in the area of Deucalionis regio, a long distance from Eridania.

Tsuneo Saheki could also report on another phenomenon which had been seen occasionally in the past. On June 4, 1937, the late Sizuo Mayeda suddenly saw an intensely bright spot which scintillated like a star, was far brighter than the polar cap, but disappeared after about 5 minutes. The place was the area of Tithonius Lacus, near the Martian equator at about longitude 95 degrees. The next instance was seen by Saheki himself on December 8, 1951. "I saw a sharp, bright, glaring spot suddenly appear on Tithonius Lacus. It was as brilliant as a sixth-magnitude star—decidedly brighter than the north polar cap—and shone with scintillation for about 5 minutes. Fading rapidly, it looked like a whitish cloudlet [5 minutes later] as large as Tithonius Lacus. [Another 5 minutes later] it was barely visible as a very faint and large white spot, and [after 40 minutes] this part of the Martian surface had returned to its normal state." Another but not so brilliant flare was seen by the same observer on July 1, 1954, at Edom Promontorium.

Wild conclusions from such observations are always drawn by those who only read about them, hardly ever by the observer himself. And Mr. Saheki's summation is a masterpiece of self-restraint and careful weighing of evidence: "We can rule out the possibility that these flares were sunlight reflected from a hypothetical water surface on Mars—their locations on the planet with respect to sun and earth preclude this. Reflection from an ice-covered mountainside is free from this objection, but cannot explain the formation of a cloud just after the disappearance of the light, as in 1951. A meteorite fall on Mars might produce both light and a cloud, but

81

meets difficulty in accounting for flare durations as long as 5 minutes. . . ." He concluded by saying that volcanic activity might account for light and cloud formation, but that the duration of the flare was too short for a volcanic eruption. Though he failed to arrive at an explanation which satisfied him, he was confident about the reality of the phenomena.

The possibility of active volcanoes on Mars, surprisingly enough, brings us to the last of the major features of the planet, the dark areas. For more than half a century practically everybody had considered them to be vegetation, but there had never been a good answer to Arrhenius's contention that they may be hygroscopic minerals changing color in the presence of moisture. That answer was finally given only a few years ago by the Estonian astronomer E. J. Öpik. Since the winds spread desert dust around, he reasoned, such areas of darkening crystals would have been covered up by yellow dust long ago, especially if the areas are, as is generally believed, at a lower level. Since they stubbornly reappear after dust storms, they must be something which has the power to break through again, and that can only be something which is alive. One might also add that, if Arrhenius were right, every observed change should consist of the gradual obliteration of one more dark area by yellow dust. But the changes actually observed for the last hundred years were usually the opposite: dark areas increased in size, extending in one direction or another, or two neighboring dark areas grew together—contact between Lacus Moeris and Syrtis major has been seen to happen.

During the opposition of 1954 a new dark area appeared in a locality where no astronomer had ever reported one before. It was observed, and photographed, by Dr. Earl C. Slipher of Lowell Observatory, working at Lamont-Hussey Observatory, Bloemfontein, South Africa. It was a large area, large even for earth, for it covered about 200,000 square miles, nearly the area of Texas. Its center was 20 degrees north of the Martian equator at latitude 235 degrees, in a sector of the Martian desert where only canals had been seen before. Dr. Slipher stated that it had the same blue-green tint of the dark areas long known, so it was evidently identical in nature with the others. And Dr. Slipher is convinced that it is vegetation.

If anybody should ask, "What kind of vegetation?" the customary answer for the last three decades has been: mosses and lichens, with special emphasis on lichens. The latter are a biologically interesting type, since lichens are really two plants living in the closest possible symbiosis. The body of the plant is a fungus, but imbedded in its threads there are cells of chlorophyll-bearing algae which can produce food by photosynthesis. Most lichens are the color of the rock on which they grow—some forms growing on rocks at often foggy seashores are bright orange—or blackish or gray. In a few cases the color depends on whether the plant is dry or moist. When dry they look gray, which is the dominant color of the fungus part of the plant, but when wet the fungus becomes translucent, the green algal cells can be indistinctly seen, and the over-all impression is a darkening with a greenish tint. In shape they are usually papery crusts, although a few of them produce a growth somewhat like a sponge in appearance. A few form hanging hairlike growths.

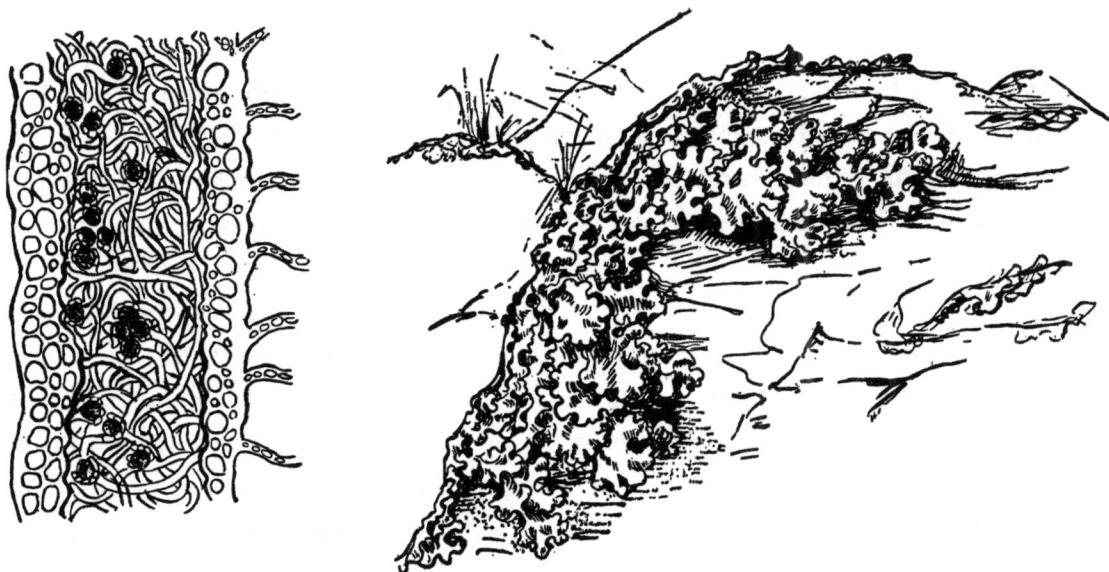

FIG. 17. Lichens. The hardy lichens of earth could probably live under the severe conditions of Mars, and it is believed that Martian vegetation is built along a similar pattern. Actually lichens are a symbiosis of two plants: algae and fungi. (*Left*) A cross section through lichen tissue, showing the round cells of the algae imbedded in the "threads" (*hyphae*) of the fungus. (*Right*) A growth of crust lichen, genus *Parmelia*, growing on rock.

The reason why the favorite answer is lichens is that they can and do live on bare rock, in any climate and at any altitude. In Greenland and in Iceland they are actually the dominant forms of plant life. In addition to everything else they extract the moisture they need from the air, having no root system. Lichens seem to be able to tolerate almost any conditions, except dry heat and air polluted by sulfur fumes. They do look as if they had been made exactly for Martian conditions, but the analogy should not be carried too far. One cannot—or at least should not—insist that Lacus Solis and Trivium Charontis are two enormous lichen patches.

83

What can really be said is that if a selection of terrestrial plants were brought to Mars the lichens are the most likely to survive. But the Martian vegetation, having evolved in a different environment, does not need to resemble our lichens in shape or even in structure; the important ability to extract moisture from the air is not restricted to lichens even on earth. In the May 19, 1950, issue of *Science* it was reported (by Dr. Edward C. Stone, *et al.*) that a mountain tree of the West Coast, the Coulter pine, is capable of sustaining itself and surviving in soil dried beyond the wilting point, by "negative respiration."

The Russian astronomer Tikhoff has devoted a good deal of study to terrestrial plants of the Arctic and of high altitudes, for the specific purpose of comparing color values with the color values of Martian vegetation. He pointed out that the vegetation of the cold tundras tends to have blue-green rather than green shades. He measured the reflection of infrared (heat) radiation from trees of the Russian north in winter. The leafless and dormant birch reflected about 55 per cent of the infrared rays, the spruce reflected only 16 per cent. All this shows that plants on earth have produced adaptations, where required by climatic necessity, and this ability must also exist, and to a higher degree, in whatever grows on Mars. But there is no telling across more than 35 million miles just what biochemical tricks have evolved under Martian conditions. In any event the dark areas, constantly breaking through covering dust, and even expanding on occasion, do not look like the "last foothold of life," as they had been described. Instead, they look rather vigorous—and it is very hard to admit the existence of large areas of vegetation and in the same breath to deny the possibility of creatures that eat those plants.

But at least some of the dark markings might be something different, as Dr. Dean B. McLaughlin of the University of Michigan suggested at the American Astronomical Society meeting in 1955. Among the changes which can be detected by a comparison of recent maps and older ones, there are a number which look to Dr. McLaughlin as if they were the results of fall-out from drifting clouds of volcanic ash. The funnel-shaped bays near the equator in the area of Sinus Sabaeus might be such wind drifts, with the volcanoes located at the points of these bays. Another feature which Dr. McLaughlin suspects of being volcanic ash is the broad Thoth-Nepenthes canal. This explanation might hold true for a few cases. But there are many features which cannot possibly be explained in this manner—the "cross" on Hellas is just one example—and the seasonal changes of color also point in another direction.

And this is the picture of Mars at midcentury: a small planet of which three-quarters is cold desert, with the rest covered with a sort of plant life that our biological knowledge cannot quite encompass. Although the air is thin, like ours 11 miles above sea level, this plant life seems to be doing well. The days are not really cold, especially near the equator, but the nights are like Arctic nights, or, better, like nights in the stratosphere. Mars is not the dead planet that Arrhenius pictured, but neither can it be inhabited by the kind of intelligent beings that many people dreamed of in 1900. If the canals actually represented the kind of engineering achievement that they were then thought to be, we can say from our present knowledge of engineering that the beings that produced them would by now have accomplished space travel.

Because we, the genus homo of earth, will set foot on Mars within a matter of decades.

5. BEYOND THE ATMOSPHERE

THE BUILDING of highways and of railroads, the planning of airports and harbors, and even the laying out of shipping lanes, on sea and in the air—all these have been called "engineering on the basis of geographical realities." Space travel, with even more right, can be called "engineering on the basis of astronomical realities." And for this reason the roots of space travel, which looks like such a modern science at first glance, go back far into the history of astronomy. When Johannes Kepler discovered the ellipticity of planetary orbits he unwittingly made a major contribution to the eventual conquest of space and the expedition to Mars, because the ships to come will travel along elliptical orbits. Likewise, when Sir Isaac Newton framed his law of universal gravitation he made an even greater contribution, because this law enables us to understand why planets and space ships must follow elliptical paths.

Other scientists, minor and major, made other contributions, small or great (but in any event unknowingly), to a science which was still in the future but which was to profit from chemistry as well as from physics, from metallurgy as well as from medicine. By the time the present century had moved well into its second decade the necessary knowledge in many fields had been amassed to the point where far-sighted physicists and mathematicians—Robert H. Goddard in the United States and Hermann Oberth in Germany—could write the first theoretical studies.

Their books were published in 1920 and 1923, respectively, and future historians will probably settle on these dates as the "beginning" of space travel. For if they should try to pick experimental accomplishments rather than the building of the theoretical foundations, it would be difficult to choose one specific event. Should they consider the first time a liquid-fuel rocket lifted itself off the ground as the beginning? If so, the year would be 1926, when one of Goddard's early rockets rose a short distance into the air. Or should they decide in favor of the first rocket shot to a height surpassing 100 miles, at which 99.9 per cent of the earth's atmosphere is below the rocket? In that case the year would be 1944, and the rocket one of the early A–4 (or V–2) rockets fired from Peenemünde. Or should they select the first successful shot of a two-stage liquid-fuel rocket? Then the date is 1949, when a two-stage rocket, consisting of a V–2 and a WAC–Corporal, took off from the White Sands Proving Ground, with the WAC–Corporal rocket reaching a peak altitude of 250 miles. Or should they decide in favor of the first manned rocket flight, with rocket-propelled aircraft? About half a dozen different dates could be quoted. Or they might accept the first artificial satellite as the real beginning, which would make the date the International Geophysical Year of 1957–1958.

Looking forward from the year 1956 it is possible to tell what the sequence of events will be, even though no dates can very well be given in advance. After the first set of artificial satellites has been fired, some of them to an altitude where they are expected still to be retarded by residual air resistance and eventually to burn up in the atmosphere, there will be satellites which have been put at a "safe" altitude and that will stay in their orbits permanently, sending the scientific information gathered by their instruments to the ground. There will probably be several satellites fairly far away which will broadcast the picture of the earth to the earth to assist meteorologists in tracing the major movements of air masses in the atmosphere. There might be other artificial satellites even farther out which are used commercially as television relay stations. According to their special purposes, these satellites will circle the earth at different distances, and very likely also in different orbital planes, some staying over the equator, others going from pole to pole, and still others in intermediate positions.

As we progress in practical experience with unmanned orbital rockets, on the one hand, and in space medicine, on the other hand, there will be airplane-like upper stages of large rockets which will carry human pilots to heights and speeds that are at present still the domain of pilotless research rockets. Then one such ship will go into an orbit around the earth temporarily, and after that the first manned space station will be built.[1] Its scientific utility will greatly exceed that of the un-

1. In two previous publications the authors of this book helped to show how manned space flight could become a reality in our lifetime. *Across the Space Frontier* (New York: The Viking Press, 1952) presented a concept of very large three-stage rocket ships capable of carrying men and substantial amounts of cargo into a permanent orbit around the earth. It demonstrated how a large wheel-shaped space station could be assembled in the orbit from parts prefabri-

cated on the ground and freighted up to the orbit with the help of such rocket ships.
Conquest of the Moon (New York: The Viking Press, 1953) went one step further and described how a full-fledged expedition to the moon could be launched from this outpost in space. It showed that a voyage to the moon and back would be a formidable undertaking, but that it was entirely feasible even with chemical rocket fuels available today.

manned artificial satellites, so it is quite likely that there will soon be other manned space stations, each designed for and devoted to a specific purpose and, like their unmanned predecessors, moving in different orbits and in different orbital planes.

Astronomical research from a space station is something every astronomer feels will solve most of the problems that puzzle him now. Our atmosphere absorbs, in round figures, two-thirds of all the radiation from space which strikes it; only about one-third reaches the ground. And that one-third is not even a representative sample of what came in, for the atmosphere absorbs some wave lengths both of visible light and of light invisible to the unaided eye in greater proportion than it does others. Located beyond the atmosphere, a telescope or a camera or a spectroscope can receive all the rays. And the "seeing" is *always* perfect in space. Many of the riddles of Mars will be solved by a telescope outside the atmosphere. But not all of them. In spite of perfect seeing and incredibly sharp and large photographs the thickness of a white spot—indicating recent snowfall—can still be debated hotly. And even if the space telescope has settled beyond any doubt that, say, Zea Lacus in the center of Hellas is an area of vegetation, there is still room for endless argument about the type of vegetation, its respirative mechanism and method of propagation.

The same limitation applies to the moon. For the last seventy years or so there has been an argument—conducted in harsh terminology on occasion—as to whether the lunar craters are impact craters, caused by the crash of large meteorites, or whether they are volcanic. Even a picture a yard wide of a small crater is not going to convince the diehards of either school of thought; both are going to insist on direct examination for either meteorite fragments or lava flows, and samples that can be analyzed chemically.

There will be expeditions. But no expedition can be made until after at least a temporary manned space station has been put together in an orbit around the earth, for the space station is, in a manner of speaking, the springboard for longer trips. The next logical step after the building of a space station is a circumnavigation of the moon, at first without landing, but timed so that the half of the moon which is forever invisible from earth (and from a space station too) is in daylight and can be mapped photographically. The ship for this trip, though probably quite small, will differ in many fundamentals from all ships built up to that moment. It will be the first of the "deep-space" ships, and from an engineering point of view

88

XVII. The first step into space: fueling a liquid-fuel, three-stage rocket which is to carry an instrumented artificial satellite into an orbit around the earth.

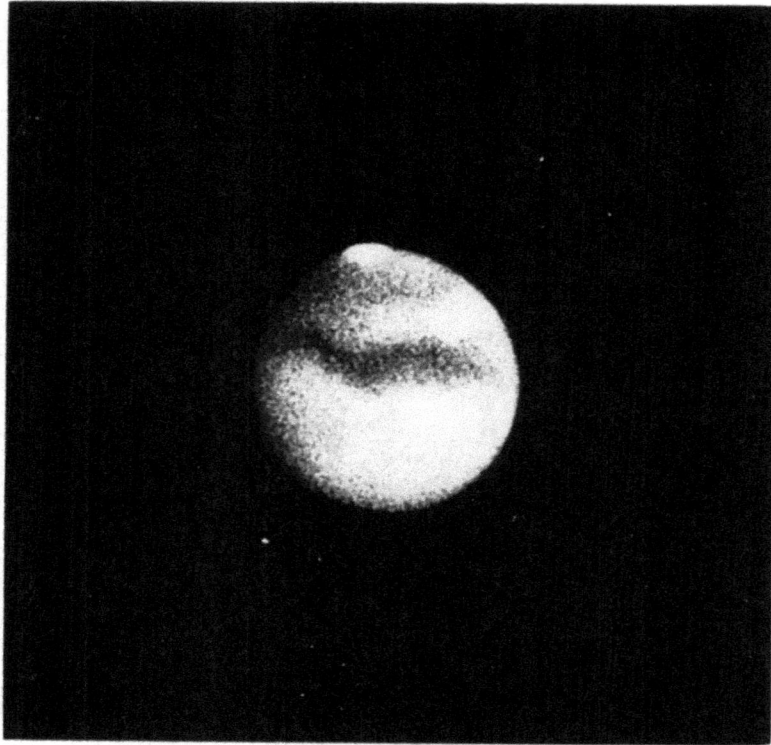

XVIII *a*. Photograph of Mars taken September 5, 1909, showing the Mare Sirenum.

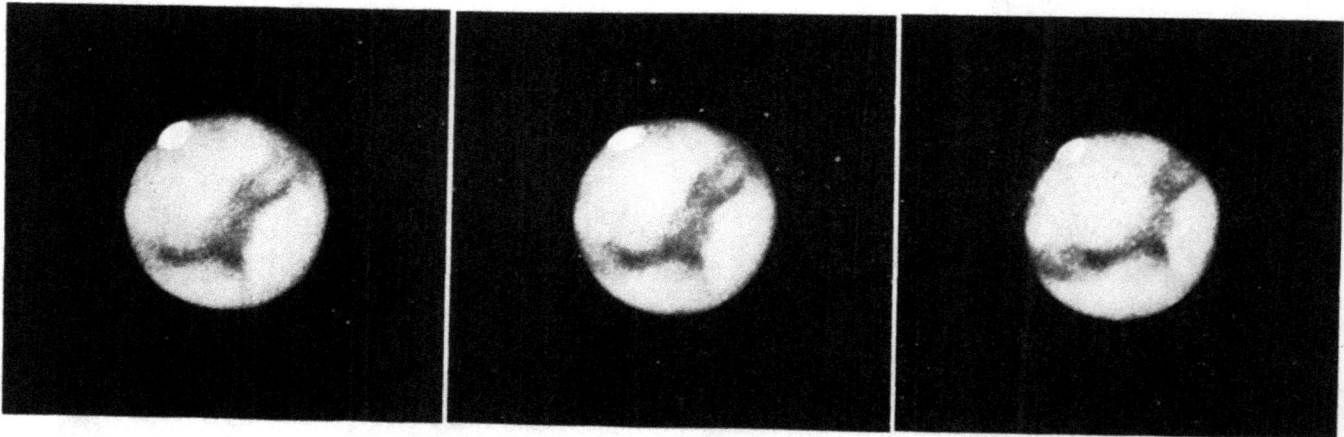

XVIII *b*. Photographs of Mars taken on September 28, 1909, showing rotation in 82 minutes.

(Courtesy Yerkes Observatory)

90

XIX. An artificial satellite, 350 miles above the 45th parallel of latitude in eastern Asia, some distance south of the city of Harbin, looking due south on Korea. The time of year is December. The southern end of the Japanese islands appears at the upper left.

XX. An unmanned instrument-carrying satellite in its orbit, passing 200 miles above the Atlantic coast. Cape Cod and Boston can be seen to the upper right; the cone of the satellite bisects Long Island; New York City and Staten Island are in the lower center; Philadelphia is near the bottom.

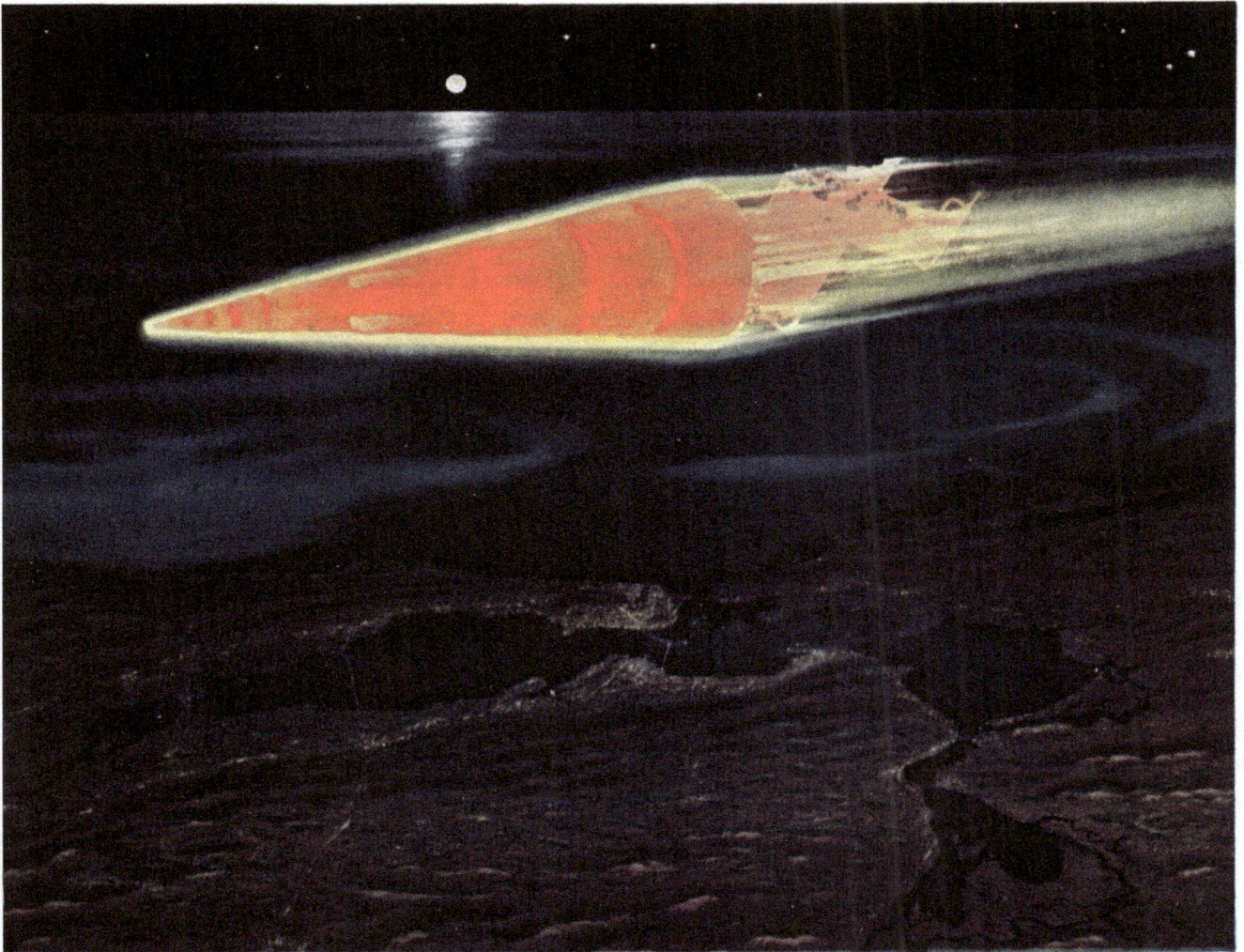

XXI. The end of an early artificial satellite, which is burning up in the atmosphere 40 miles above the San Francisco Bay area, looking 15 degrees south of due west. The re-entry of such satellites into the earth's atmosphere and their disintegration are valuable parts of the experiment, providing information about some aspects of atmosphere re-entry that cannot be obtained otherwise.

93

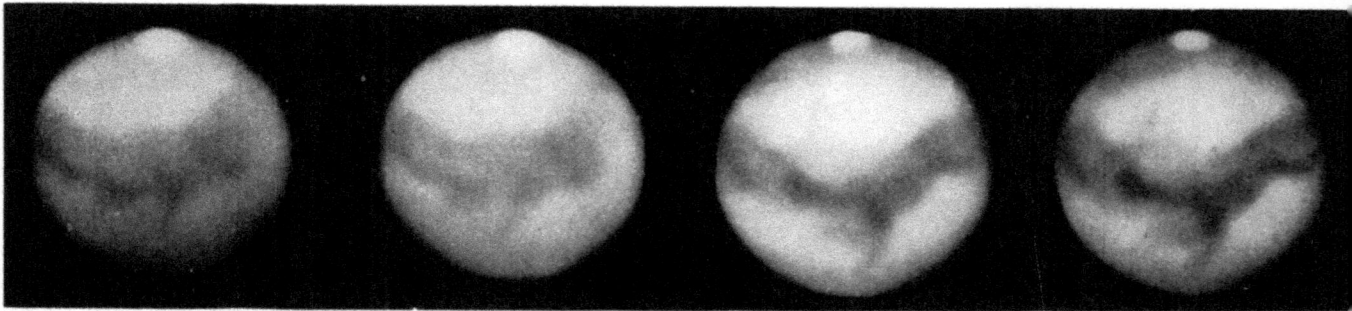

| 1 | 2 | 3 | 4 |

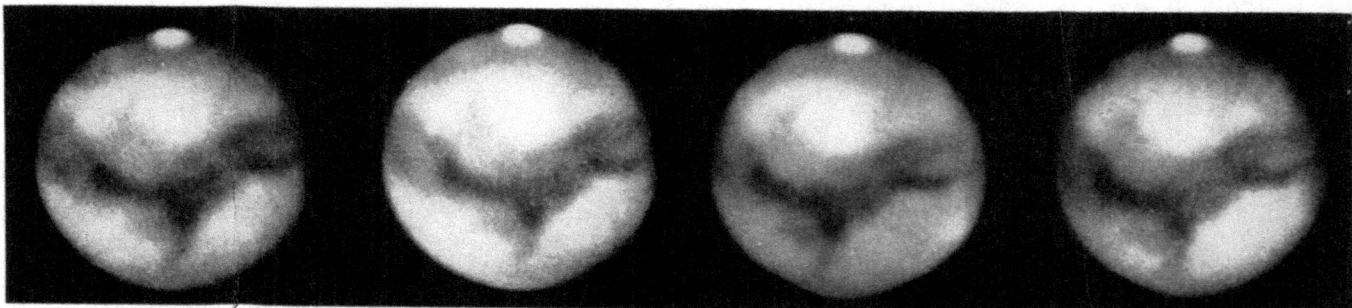

| 5 | 6 | 7 | 8 |

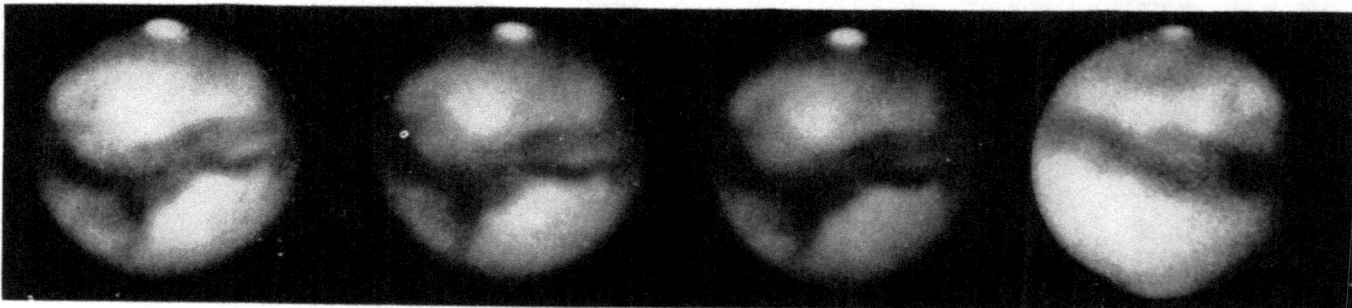

| 9 | 10 | 11 | 12 |

XXII. Mars in 1909. Composite photographs, made by superimposing photographically three to five negatives taken in rapid succession. This technique eliminates accidental detail and strengthens the visible, or rather the photographable, features. Nos. 1 and 2, taken September 25, show a veiled atmosphere; on Nos. 3 and 4, taken September 29, the atmosphere is clear. Nos. 5 to 8 all show Syrtis major; all were taken on September 29, Nos. 7 and 8 about 40 minutes later than Nos. 5 and 6. Nos. 9, 10, and 11 also show Syrtis major; these were taken on the same day as Nos. 5 to 8, but about half an hour after Nos. 7 and 8. No. 12, taken on October 6, shows the area of the Mare Cimmerium.

(Courtesy Yerkes Observatory)

XXIII. Mars in 1909.

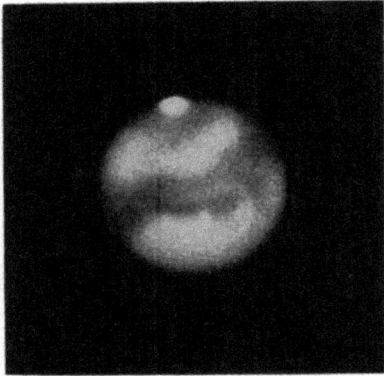

Showing area of
Margaritifer Sinus,
September 24, 1909.

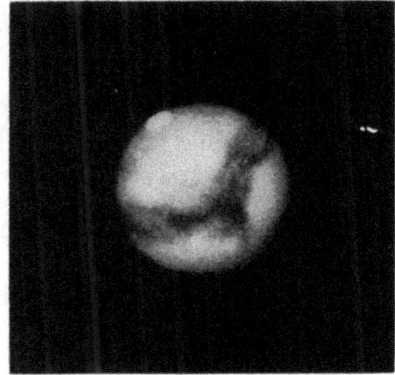

Showing area of
Syrtis major,
September 28, 1909.

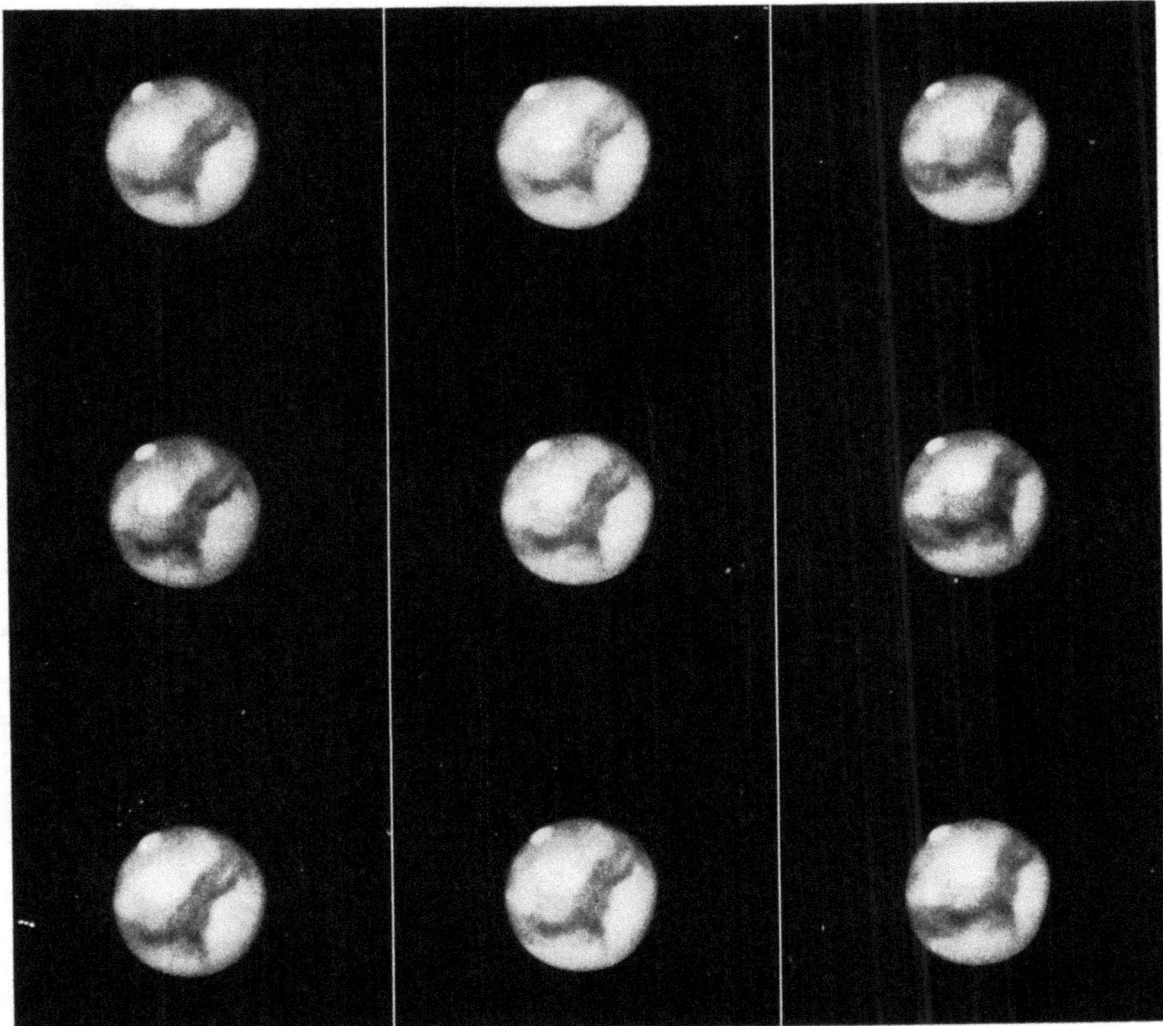

Photographs of Mars showing Syrtis major; taken on September 28, 1909, at about hourly intervals to show rotation.

(Courtesy Yerkes Observatory)

XXIV. Assembling the ships for the Mars expedition, 1075 miles above the west coast of South America. The group of islands visible at the lower left is the Galápagos.

96

deep-space ships which do not take off from the ground present fewer problems. For example, any take-off from the ground requires a rocket thrust considerably greater than the ship's take-off weight, simply to enable it to lift itself off the ground. For departure from an orbit around the earth this relationship between rocket thrust and ship's weight is not required, because the weight of the rocket ship is fully compensated for by the centrifugal force caused by the orbital motion. A deep-space ship can, and of course will, have rocket engines that are considerably weaker and lighter than those of any ship designed to take off from the ground.

Moreover the ship does not need streamlining of any kind, since it will never enter any atmosphere. It can have any shape that is convenient for structural or other reasons. But though it is not meant to go through any atmosphere, it will still be built on the ground, like the space station, and transported to an orbit around the earth in the streamlined ships which carried the space station into its orbit piecemeal. And it will have to be reassembled in space; the space station is going to be the place where the assembly crew eats and sleeps.

This trip around the moon without landing will be a very important test. The small crew of this ship will be used to space, of course; every member of it will have made the trip to the space station and back repeatedly. And every one of them will have spent some time in the space station. But the trip to the space station from the ground takes only about one hour, and living in the relatively spacious quarters of the space station is not the same as living in the small cabin of a space ship. The 10-day trip around the moon will be a fine service test for both men and machines and will furnish practical experience in the difficult art of space navigation and the use of its tools: radio and radar tracking methods, as well as the spaceman's equivalents of present-day aerial and naval astronavigation techniques. Some years after the first flight around the moon, an expedition consisting of two or more larger "deep-space" ships might attempt a landing on the moon. But even after this expedition has returned successfully from a 4- or 6-week exploration of the moon, we shall have only a faint understanding of the problems connected with man's retaining his spiritual and mental and physical health while traveling for months and months through the emptiness separating the earth from Mars. The expedition to Mars should be considered the ultimate achievement of a gradual and often painful step-by-step development of manned space flight which may take many decades to accomplish.

THE EXPLORATION OF MARS

Technological prophecy spanning a time interval of decades is handicapped by the rapid progress of the natural sciences and the likelihood of developments of fundamentally new methods. It is entirely possible, for example, that within a decade or so successful tests with some sort of nuclear rocket power system might be accomplished. But the chances are that nuclear rocket propulsion systems will find their first application not in ground-launched rockets but in deep-space rocket ships.

There are many good reasons for believing this. Deep-space ships can operate for longer periods of time with less thrust. This reduces power requirements, heat transfer rates, and the weight of the reactor. Outside the atmosphere the heavy radiation shield for the protection of the crew can be limited to line-of-sight shielding between reactor and cabin, also called "shadow shielding." Nor is there any problem of radioactive contamination of the launching area in the vacuum of outer space.

Therefore, it seems probable that chemical rockets will still be used for flights from the earth's surface to the orbit of departure, even after a nuclear rocket system has been developed. And it is equally probable that the fundamental concept of subdividing an interplanetary expedition into an orbital supply operation, the interplanetary voyage proper, and a landing operation—each with separate and different vehicles—will still be adhered to.

Many new scientific discoveries will be made before the time will be ripe for a voyage to Mars. Many inventions of which we have no conception today—and not in the field of propulsion only—are going to be at the disposal of the engineers in charge of the actual design of the Mars ships. Nevertheless, the laws of astronomy and of mechanics permit us to analyze the present technical requirements for an expedition to Mars, and it is exciting as well as instructive to translate the results of such an analysis into engineering solutions based on our present technical know-how. While the resulting designs may be a far cry from what Mars ships some thirty or fifty years from now will actually look like, this approach will serve a worthwhile purpose. If we can show how a Mars ship could conceivably be built on the basis of what we know now, we can safely deduce that actual designs of the future can only be superior. Only by stubborn adherence to engineering solutions based exclusively on scientific knowledge available today, and by strict avoidance of any specu-

98

lations concerning future discoveries, can we bring proof that this fabulous venture is fundamentally feasible.[2]

The basic "astronomical reality" guiding all thinking about an expedition to Mars can be read from Fig. 5 in Chapter 2. It is the fact that all planets move around the sun in the same direction and in about the same plane. As Kepler realized, the orbital velocity of the planets is the higher the closer they are to the sun. What he did not know is that the nearly circular orbits of the major planets may be described as demonstrations of a neat balance between the centrifugal force produced by the orbital velocity of a planet and the gravitational pull of the sun for that distance. Our earth, at an average distance from the sun of 92.9 million miles, needs its orbital velocity to stay in its orbit; the velocity is 18.52 miles per second. If the sun suddenly stopped having a gravitational field, the earth would, at that instant, begin to move with that velocity along a straight line which would be a tangent to its former orbit.

Mars, being farther away from the sun—the average distance is 141.5 million miles—needs a lesser velocity to stay in its orbit; the average is 14.98 miles per second. Let us assume that it were possible to increase the orbital velocity of the earth at some arbitrary point of its orbit. Earth would then be "too fast" for the present orbit; the sun's gravitational attraction would be insufficient to hold the earth at a distance of 92.9 million miles, and it would recede from the sun in a new elliptical orbit. If we could increase the earth's velocity by 1.88 miles per second to a total of 20.40 miles per second the aphelion of the new orbit would touch the orbit of Mars. The new orbit would be an ellipse with the aphelion in the orbit of Mars and the perihelion in the former orbit of earth.

2. In 1953, one of the authors of the present volume published a scientific study investigating the feasibility of a voyage to Mars with chemical rocket fuels (*The Mars Project*, by Dr. Wernher von Braun, Urbana, Ill.: University of Illinois Press, 1953). The study dealt with a hypothetical expedition of no less than seventy men traveling in ten rocket ships, and it arrived at the conclusion that such an expedition was feasible, theoretically at least, if only someone could be found to foot a fuel bill equivalent to that of ten Berlin Airlifts. The backbone of the undertaking was to be a fleet of three-stage orbital rocket ships, identical with those described in *Across the Space Frontier*. In 950 ferry flights, shuttling back and forth between the ground base and the orbit of departure, these rocket ships were to carry the prefabricated parts of the Mars ships, their fuel, the expeditionary equipment, and the crews to the orbit, where the Mars vessels would be reassembled and whence they would finally depart for the Red Planet.

The following may be considered a revision of this study. Although it envisions an expedition of only twelve men traveling in two ships, the total propellant requirement—a good yardstick for the over-all logistic effort—is only 10 per cent of that found in *The Mars Project*. This enormous saving is due solely to a superior over-all plan, for the specific assumptions made with regard to rocket engine performance and construction weight factors have *not* been altered.

THE EXPLORATION OF MARS

While we cannot increase the orbital velocity of the entire earth, this could be done with a minute portion of it. If a rocket ship could leave the gravitational field of the earth with a speed of 1.88 miles per second, in the direction of the earth's own orbital motion around the sun, it would become that minute portion.

Of course, the rocket ship would first have to break away from the gravitational field of the earth, if it is to escape into open interplanetary space with a "residual speed" of 1.88 miles per second. But by departing from an orbit around the earth, instead of taking off from the surface, we can ease this task.

Three factors account for this. First, the Mars-bound ship would then depart from an altitude where the earth's gravitational pull is markedly weaker than on the ground. Second, the ship would not need to fight air drag because the flight would begin above the atmosphere. Third, and most important, the orbital speed would give the Mars-bound ship a running start. For example, if an orbit of departure of an altitude of 1075 miles were chosen (corresponding to a period of revolution around the earth of exactly 2 hours), the initial orbital speed of the Mars ship would be 4.40 miles per second. If the Mars ship were accelerated by rocket power to a speed of 5.99 miles per second—just 1.59 miles per second more—

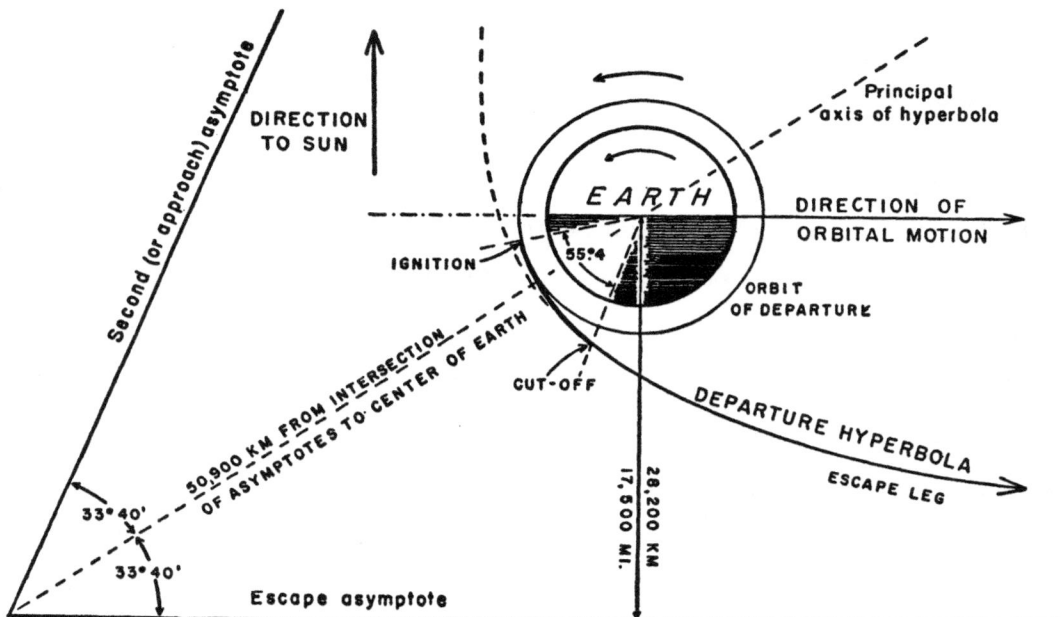

FIG. 18. Maneuver 1—departure from an orbit around earth. The actual departure maneuver lasts from the point marked "ignition" to the point marked "cut-off." The increase in speed resulting from the firing of the rocket engines causes the two ships to leave their circular assembly orbit and to move along the escape leg of the hyperbola.

while climbing to a cut-off altitude of 1965 miles, it would leave the earth's gravitational field with that required residual speed of 1.88 miles per second. This period of acceleration, the departure maneuver, is the first of a total of four power maneuvers of the contemplated voyage.

It is possible to time and to execute this departure maneuver in such a manner that the ship will be moving exactly in the direction of the orbital motion of the earth when, at a distance of a million miles or so ahead of the earth, the earth's gravitational field will have dwindled to such an infinitesimal value that one can say the earth's gravity has been left behind. Henceforth the ship's motion is influenced only by its momentum and by the gravitational field of the sun. Without need for additional power application the ship will follow an elliptical path on which, 260 days later, it will touch the orbit of Mars.

Perhaps the most noteworthy feature of such a voyage through the solar system is that the rocket ship, for reasons of fuel economy, does *not* take the shortest route to Mars. Indeed, by coasting halfway around the sun, it takes a very, very long route. We just have to get accustomed to the fact that on an interplanetary journey one does not travel "as the crow flies." Being temporarily a tiny, life-supporting heavenly body on its own, the ship rather coasts through the solar system like a comet. Only by making fullest use of the initial speed provided by the earth's own orbital motion around the sun can the ship swing out to the Mars orbit with a minimum of fuel expenditure. The actual mileage covered during such a one-way voyage to Mars therefore is far in excess of the 35 million miles that lie between earth and Mars during most favorable oppositions. It is no less than 735 million miles! As regards the time required for this comet-like coasting from orbit to orbit, it is rather obvious that this must be longer than half an earth year (365:2 = 182.5 days) and shorter than half a Mars year (687:2 = 343.5 days). A simple calculation yields a duration of 260 days, or a little over eight months, for the one-way trip (see Fig. 19).

These 260 days must be carefully taken into account in choosing the departure date. After all, we are not interested in intercepting the orbit of Mars at the aphelion of the voyaging ellipse: what we want is to make sure that the planet itself will be at that point of its orbit when the ship arrives. Obviously the ship must depart from the vicinity of the earth at a time when Mars is in such a position that it will need 260 days to get to the rendezvous point too. The rendezvous point, of course, is the aphelion of the voyaging ellipse.

101

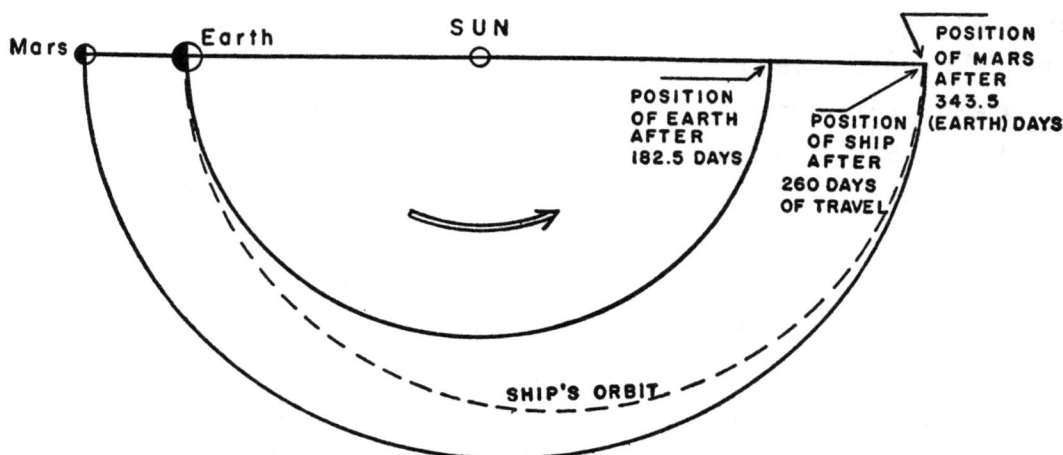

FIG. 19. The duration of the flight along the half-ellipse connecting the orbits of earth and Mars is longer than half an earth year but shorter than half a Mars year. The actual figure is 260 days.

The position of the plane of the voyaging ellipse is the next point to be considered. Prior to the departure maneuver the ship should circle earth either in the plane of the ecliptic or, to be precise, in the slightly diverging plane of the Martian orbit (see Fig. 21). The fact that the ship must circle earth for some time in the departure orbit unfortunately introduces a complication. Because of the oblateness of the earth the plane of the orbit of the circling ship will be subjected to certain periodic changes.[3] But since these changes can be calculated and predicted, they do not constitute an unforeseeable complication, but merely a fact which has to be taken into account to be sure that the plane of the departure orbit is properly aligned on the date of departure, which is determined by the position of Mars.

During the 260 days of unpowered coasting through the half-ellipse, the Mars-bound ship will be continuously increasing its distance from the sun, which means that it will be coasting "uphill" against the steady pull of solar gravity. As a result, it will gradually lose some of the initial speed of 20.40 miles per second with which it entered its circumsolar flight path at the perihelion—the point of the cruising ellipse closest to the sun. When the ship arrives at the aphelion, tangential to Mars' orbit, the speed will be down to 13.39 miles per second, and Mars will overtake the ship from astern with an excess velocity, relative to the ship, of 1.59 miles per second. If the ship, when arriving at the aphelion, were exactly in Mars' orbit, it would be caught by Mars' gravity and crash upon Martian soil.

3. The so-called regression of the nodes; see *Conquest of the Moon*, pp. 22 ff.

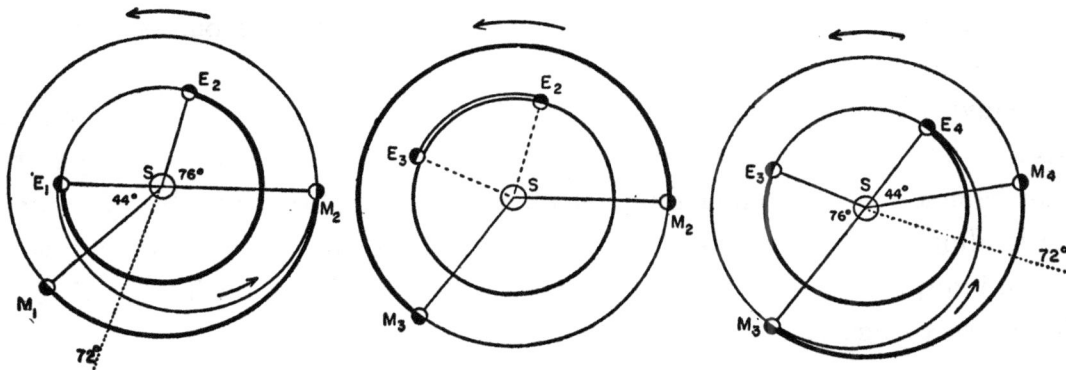

FIG. 20. PLANETARY MOVEMENTS DURING THE MARS EXPEDITION

The Mars-bound voyage (*left*). At the beginning of the voyage the slower-moving Mars must be 44 degrees of arc ahead of earth, the two planets occupying the positions E_1 and M_1. On the 73rd day of the voyage both the expedition and the earth have moved through an angle of 72 degrees so that sun, earth, and ships form a straight line, causing a "transit" of the earth and moon across the disk of the sun. After 260 days ship and target planet meet at M_2 but at that time the earth is 76 degrees ahead of Mars at E_2.

The waiting period (*center*). During the waiting period of 449 days the planet Mars moves from M_2 to M_3, through an arc of 235 degrees. The earth, moving very nearly 1 degree per day, has completed a full revolution around the sun, plus an arc of 83 degrees, a total of 443 degrees. At the end of the waiting period the earth, therefore, is at E_3 which, as seen from the sun, is 76 degrees behind Mars. The return voyage home cannot begin until the two planets are in this position relative to each other.

The homebound voyage (*right*). The return voyage begins when Mars is at M_3. While the ship moves through 180 degrees during the 260-day trip, the earth moves 180 plus 76 degrees, so that the two will meet at E_4. When the ship is still 72 degrees behind E_4—on the 187th day of the trip—another transit of earth and moon across the sun's disk will take place. During the time required for the voyage, the planet Mars moves through an arc of 136 degrees and will be at M_4, 44 degrees behind the earth.

But we can locate the aphelion of the circumsolar voyaging ellipse just a few thousand miles *inside* of Mars' orbit. Mars will then pull the ship down with ever-increasing speed through a hyperbolic arc (see Fig. 22). At the vertex of this hyperbola the ship would be closest to the Martian surface—and it would escape Martian gravity again for all eternity on the second leg of this grazing sweep, if it were not slowed down. By use of the ship's rocket motors, acting as brakes, the speed can be reduced appropriately just before it reaches the vertex, and the ship can be induced into a circular orbital path around Mars. If we choose a distance of the aphelion of 5470 miles this side of Mars' orbit, the altitude of the vertex of the approach hyperbola will be 620 miles above the Martian surface, and this will also be the approximate altitude of the final circum-Martian orbit of the ship. The rocket motors must impart a velocity reduction near the vertex of 1.25 miles per

THE EXPLORATION OF MARS

second during this "capture maneuver," and this will be the second power maneuver for which rocket fuel must be provided.

Having become an artificial satellite of Mars, the ship may now remain in this orbit as long as desired. From this position the descent to the Martian surface must be made in a special landing vehicle which must also be capable of returning the landing party to the orbiting ship. This phase of the voyage, which involves not an interplanetary but a "local" transportation problem, is dealt with later; now the return voyage to earth must be considered.

The departure from the circum-Martian orbit is the third power maneuver for which rocket power and fuel are required. The velocity change required for leaving the orbit will be the same as for the capture maneuver—namely, 1.25 miles per second. The exact instant of the departure (in terms of minutes and seconds) is determined by the condition that the ship, when leaving the last traces of Martian gravity at a speed of 1.59 miles per second, must be coasting in a direction exactly opposite to Mars' own orbital motion around the sun. But 14.98 (the speed of Mars) minus 1.59 still leaves 13.39 miles per second. Through another half-ellipse (which may be looked upon as a natural continuation of the Mars-bound half-ellipse that was interrupted by the capture maneuver) the ship will then coast back to the earth orbit.

The time of earth-bound departure from the circum-Martian orbit (and now we are talking in terms of days) is tied to a condition similar to that for the departure from the earth orbit: we must be sure that the earth will be at the rendezvous point when the ship passes through the perihelion of the return ellipse. Since the ship again requires 260 days for the return voyage, the earth-bound departure must take place on a day when the earth itself is 260 days in front of the anticipated rendezvous point.

The ship's distance from the sun is continually decreasing during its unpowered coasting through the half-ellipse of return. The ship is now going "downhill" and will therefore pick up speed. As a result it will sweep tangentially into the earth orbit at a speed of 20.40 miles per second, overtaking the earth from the rear with an excess speed of 1.88 miles per second. The ship must now be maneuvered so that it will approach the earth from a point many thousands of miles outside the earth orbit, lest it be pulled straight down and crash vertically into the earth's atmosphere. If we select a displacement between the perihelion and the earth's orbit of 79,000 miles, the ensuing hyperbolic fall into the earth's gravitational field will lead through the vertex at a speed of 2.65 miles per second (see Fig. 23). In the fourth and final maneuver, using its rocket engines again as brakes, the ship will reduce this speed by 1.33 to 1.32 miles per second, thus settling in an orbit around the earth of a radius of 56,000 miles.

To summarize, the problem of reaching Mars, starting out from an orbit around the earth with subsequent return into a similar (but by no means identical) orbit, is subdivided into four main power maneuvers, as follows:

(1) Departure from an orbit around the earth at a comparatively short distance from the surface. This results in a trip along half of a Keplerian ellipse with aphelion near the Martian orbit.

(2) Capture by Mars in an orbit around the planet; the approach to the planet is a hyperbola around the planet, changed into a near-circular orbit by rocket braking.

(3) Departure from the orbit around Mars, resulting in a trip along the second half of the Keplerian ellipse with perihelion near the orbit of earth.

(4) Capture by earth in an orbit around earth; the approach is also a hyperbola which is changed to a near-circular orbit around the earth, but at a much greater distance from the surface than was the departure orbit.

In addition to these four main propulsion periods, several additional short corrective bursts of rocket thrust will be needed. It is rather obvious that the approach to the two capture maneuvers especially calls for meticulous navigation. Take the capture in the Mars orbit, for instance. If the ship is to settle in an orbit 620 miles above the surface of Mars, it must be positioned in such a manner that it is drawn into the Martian field of gravity from a point 5470 miles on this side of Mars' orbit. Now the distance between this initial contact point with the Martian field of gravity and the perihelion (where the ship left the earth) is around 234 million miles (the sum of the distances of earth and of Mars from the sun), for the ship had to coast halfway around the sun to get there (see Fig. 21). It is clearly too much to expect that the ship, after 260 days of unpowered flight, will arrive exactly 5470 miles inside of the Martian orbit. This can be accomplished only by a corrective power application which in all probability would be made a few days prior to capture, when Mars is catching up with the ship. It can be calculated in advance just where Mars should be relative to the ship at a given moment and what the relative distance between ship and planet should be. The navigator can establish his positions by determining Mars' location against the background of fixed

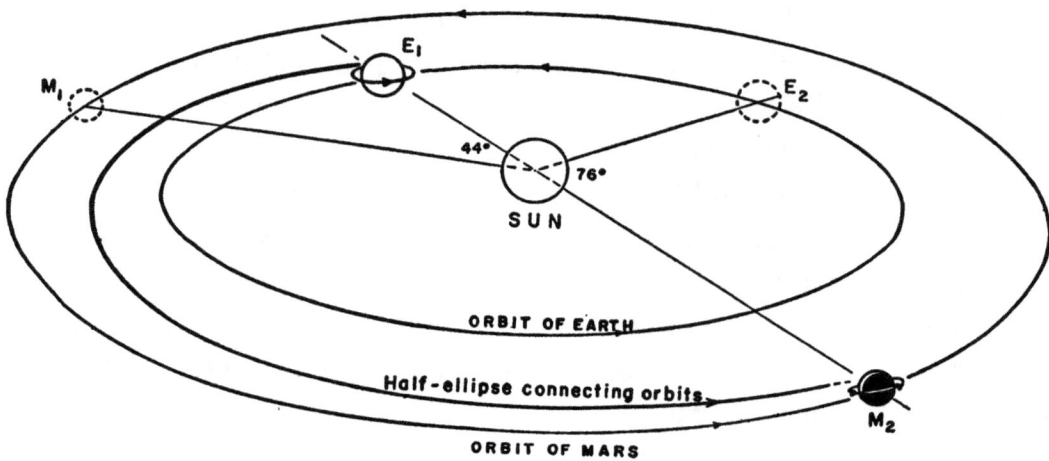

FIG. 21. Positions of the two planets during the trip to Mars. When the ships arrive near the orbit of Mars in the aphelion point of their voyaging ellipse, their velocity is reduced to 13.39 miles per second, and since Mars moves at the average velocity of 14.98 miles per second, it is 1.59 miles per second faster than the two ships. But the velocity of the ships is then increased because of the gravitational attraction of the planet; a portion of this excess speed will have to be nullified by rocket action. The over-all result will be a circular orbit around Mars, in which the ships travel around the planet in the same direction as the moons of Mars but inside the orbit of the inner moon. (Compare Figs. 20 and 22.)

stars, and he can find the distance between ship and planet by measuring the planet's apparent diameter—with a cross check by radar—and can then recommend to the captain just what action should be taken to make the ship's actual orbit agree with the precalculated conditions.

The velocity changes of the ship required for the four main power applications, plus the probable amount of velocity changes for corrections, determine the total amount of rocket propellants that must be provided. It is important to keep in mind that the propellants to be used for the later maneuvers represent ballast during the earlier ones. Therefore the amounts of fuel needed even for equal velocity changes are not equal; a given velocity change during the early part of the voyage consumes far more fuel than does an equal change during the last portion of the trip.

Before we can determine the quantity of propellants actually needed, we must make a master plan for the expedition. First, we must find out how long the entire voyage will last. Next we have to decide how many people shall go along. We have to determine what kind of supplies the explorers need, and in what quantities. This supply business is a very important consideration, which must not be regarded

106

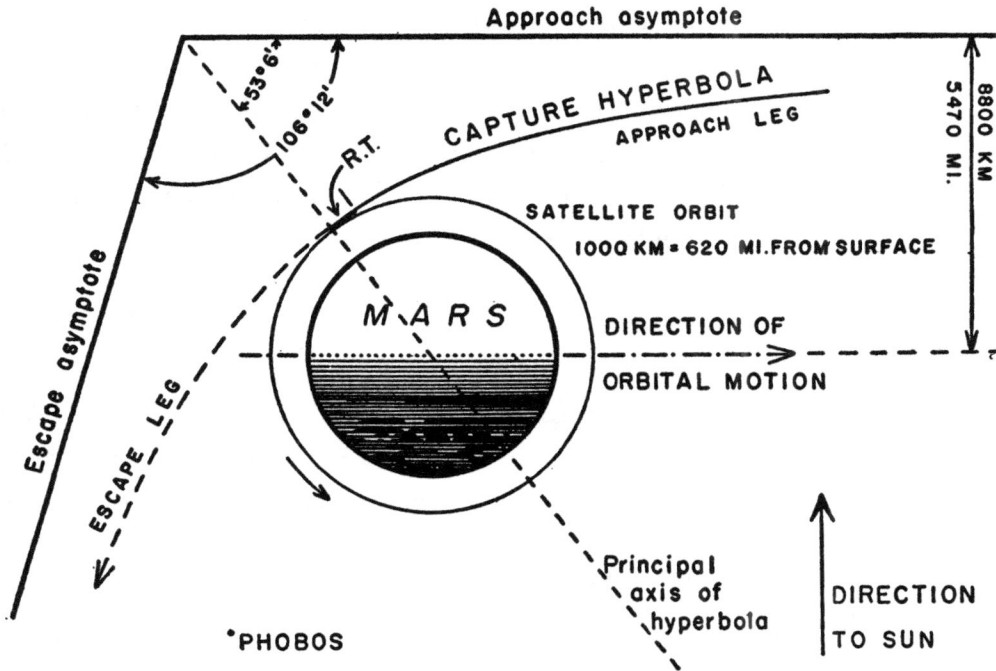

FIG. 22. Maneuver 2—capture of ships by Mars. The gravitational field of Mars pulls the two ships from the elliptical path around the sun and forces them into a hyperbola. Retarding the speed of the ships on the approach leg of this hyperbola changes the movement along the hyperbola into a circular orbit around Mars 620 miles above its surface. The short section labeled R.T., on the approach leg of the hyperbola, is the retardation track, during which the ships are slowed down. All jettisoned matter, which is not retarded, continues along the escape leg.

as a side issue in an undertaking of such magnitude. After all, a voyage across the space separating two planets is not like an automobile trip, during which the traveler can breathe the air of the state through which he is traveling and where he can have his car serviced at any filling station. Nor can we expect Martians to greet the explorers with refreshments. We have to provide the crew with virtually everything for the entire duration of their absence from the earth—air to breathe, food and drinking water, repair tools, spare parts, heatable and pressurized quarters for the stay on the cold Martian plains, surface vehicles and fuel for them, down to such prosaic items as a washing machine and a pencil sharpener. The expedition will need a two-way radio station capable of crossing the several hundred million miles which at times will separate the lonely travelers from the earth. And it will require a powerful telescope to enable them to survey the Red Planet from the vantage point of the circum-Martian orbit to determine a suitable landing site prior to the descent.

107

Some of the equipment is needed only for the orbit-to-orbit portions of the undertaking; some must be taken down to Mars' surface. Some of the supplies will be consumed on the Mars-bound flight, others on Mars itself, and the remainder must be stored against the 260-day trip home. Much equipment will be abandoned en route, since hauling it all the way back to earth wouldn't pay. Thus the payload requirements for the various phases of the trip will continually decrease. With a sound master plan, aiming at highest fuel economy, the expedition should return into the earth orbit with only a minimum of supplies and equipment left.

Let us first add up the figures for the total duration of the voyage. We have seen that it will take 260 days to get to Mars and another 260 days to return to earth. But in discussing the rendezvous problem we have also seen that the expedition can depart from the earth only on a day when Mars assumes a certain angular position in its orbit relative to the earth. Likewise, the travelers can leave Mars only when the earth arrives at a certain point in its orbit. These two relative positions of earth and Mars in their orbits, which determine the "permissible departure dates" for the Mars-bound and the earth-bound voyages, are predictable for years and even centuries in advance, in a similar manner to and with as much accuracy as the prediction of two consecutive solar eclipses. But the dates also contain a hidden instruction on how long a ship must wait in the circum-Martian orbit for the next permissible date for a return voyage. This "waiting time," during which the descent to the Martian surface, the ground exploration, and the re-ascent to the circum-Martian orbit must be accomplished, lasts 449 days, almost one year and three months (see Fig. 20, center). Add to this twice 260 days for the Mars-bound and the earth-bound voyages and you arrive at 2 years and 239 days for the total duration of the venture.

In order to keep the costs for the undertaking to a minimum, the expedition shall be limited to twelve men. Their total weight, 2640 pounds, will be an almost negligible item in the payload list. So will be the allowance for personal baggage, 220 pounds per man. Each member of the expedition will require 2.72 pounds of oxygen per 24-hour period for breathing; this will be taken along in liquefied form. As regards food, a consumption of 2.64 pounds per man per day, which is about Army standard, seems reasonable. Drinking water will amount to 4.4 pounds per man per day. It is not necessary to provide a water supply for washing, cleaning, and similar purposes, because this water supply—called "utility water"—will be produced automatically as time goes on. Of the 4.4 pounds of liquids consumed per day by every man a large percentage—on the order of 3½ pounds—goes into the atmosphere of the ship by way of exhalation and skin evaporation. This water must be extracted by the air-conditioning and air-purifying units to keep the humidity at a comfortable level and it thus becomes available for utility purposes, after thorough sterilization, of course. In an emergency this water, since it has been sterilized, could even be re-used for drinking—by the time the Mars expedition can be planned in detail much information along these lines will be available from the operation of the space station.

In any event a rather large supply of un-needed utility water is likely to accumulate in the periods between power maneuvers. This surplus water, plus toilet accumulation and garbage, will be jettisoned prior to each power maneuver.

The allowance for books, tables, and navigational aides is 1540 pounds; for ship's tools and spares, 4200 pounds; and 3600 pounds for the telescope for scanning the Martian surface from the orbit. Another 4300 pounds are set aside for sounding missiles to study Mars' atmosphere from the circum-Martian orbit. With 4400 pounds it should be possible to build a complete two-way radio station, including antenna, for communication with the earth, for a meager 10 kilowatts of transmitting power would provide ample power to bridge those vast interplanetary distances.

The heavyweight on the payload list, however, is the landing craft needed for the descent from the circum-Martian orbit to the surface of Mars. In essence, this is a large airplane capable of performing a long glide through the thin Martian atmosphere. The maneuver itself will greatly resemble the return of a winged rocket from the space station to the ground. A short burst of rocket power, with the exhaust nozzles aimed in the direction of the movement, will reduce the speed of the landing craft circling Mars so that it will enter the upper layers of the Martian atmosphere along a tangential line. Being retarded some more by air resistance, it will sink deeper into the atmosphere and perform a very long glide along a very shallow approach path. A few hundred feet above the ground the special landing gear will be lowered, and finally the large glider will land, airplane fashion, on the sands of the Martian plains.

For two reasons, the landing craft must be rather large and heavy. In the first place, it has to take enough cargo to sustain the landing party for the entire duration of their stay on Martian soil. If we assume that nine of the twelve men will descend to the surface and spend 400 of the 449 days' "waiting time" on Mars it-self, the cargo for oxygen, water, and food alone will be 18.7 tons. We must provide the explorers with a heatable, collapsible tent, inflated by an artificial atmosphere within, to enable them occasionally to get out of their pressurized suits and to protect them against the bitter cold of the Martian nights. Furthermore, we have to furnish them with research gear and a minimum of surface transportation to enable them to do a little more on Mars than just to walk importantly around their landed glider. For all this and more we will allow another 35 tons of cargo.

But there is a second and even more important reason for the great weight of the landing craft. It must carry enough fuel down to Mars to be able to return under its own power to the circum-Martian orbit. Fortunately, Mars' gravitational field is much weaker than that of the earth, and therefore this return can be accomplished in a single-stage rocket flight. Sometime prior to the return ascent, the glider will shed its wings, cargo bin, and landing gear (see Plates xxxi and xxxvi-xxxvii). With the help of the winch-equipped surface vehicles the wingless hull will then be erected into a vertical position. For the return flight, it will launch itself, rocket fashion, back to the circum-Martian orbit. It will carry only the nine explorers, plus an allowance of 5.5 tons for research specimens to be collected on Mars. All equipment used on the surface will be left behind.

All these requirements result in a landing craft of considerable dimensions and weight. When leaving the circum-Martian orbit, the craft will weigh 177 tons. It will spend 12 tons of fuel to swing itself into a landing ellipse whose lowest point will "graze" the Martian atmosphere at an altitude of 96.5 miles. With a weight of 165 tons it will land on Mars' surface.[4] Cargoes left behind and supplies used up on Mars will amount to 54 tons. Wings, cargo bin, and landing gear shed prior to the return take-off weigh 35 tons. Thus the weight of the craft will be reduced to 76 tons when the return flight begins. Of this, 62.2 tons of fuel will be burned up during the ascent, so that the small wingless hull of the once mighty and heavy glider will return into the circum-Martian orbit with a weight of only 13.8 tons.

For the interplanetary portion of the Mars-bound voyage, however, every one of the 177 tons of the fully loaded and fueled glider constitutes payload. A deep-space ship of truly gigantic dimensions would be needed to carry the heavy land-ing craft to the circum-Martian orbit, and still have enough fuel left for the return voyage to earth. But we can simplify our task greatly by using two deep-space

4. All these weights are "earth tons." On Mars the landing weight will be only 63 tons.

ships. One, the "passenger ship," is designed for the entire orbit-to-orbit round trip and for no more payload capacity than is absolutely necessary. The other, the "cargo ship," is designed for the one-way trip only. In lieu of the weight for the return fuel (that is, the fuel for the last two main power maneuvers and the corrective maneuvers during the return flight), it carries the fully loaded landing craft, and, in addition, all supplies and extra equipment needed up to the day of departure from the circum-Martian orbit. With this simple strategem the dimensions and weight of each ship can be cut down to manageable proportions.

Each of the two Mars ships, ready for departure, will weigh 1870 tons. Initially, each ship is powered with twelve rocket engines, using hydrazine as fuel and nitric acid as oxidizer. The twelve engines develop a combined thrust of 396 tons. The first power maneuver, the departure from the earth orbit, lasts 948 seconds. During this time the twelve engines gulp up 1370 tons of propellants, or about 73 per cent of the ship's initial weight. The initial acceleration is low—just about one-fifth of normal gravity. As a result of the weight loss caused by fuel consumption during the first power maneuver, the acceleration gradually climbs to about 7/10 g. By the moment of power cut-off, both ships have climbed from their original orbital altitude of 1075 miles to 1965 miles, and their speed has increased from their orbital velocity of 4.40 miles per second to 5.99. This first power maneuver, properly timed and guided, puts the two ships directly in the unpowered voyaging ellipse to Mars. With their weights reduced to about 500 tons, they will faithfully follow this prescribed elliptical path like "two comets in formation flight" until, 260 days later, their flight paths will approach the Martian orbit.

Prior to the capture maneuver, but after all corrective power maneuvers have been completed, each ship will jettison the four empty spherical containers which held propellants for the departure maneuver, along with all the refuse accumulated during the 260 days' voyage, in order to save fuel. Both ships will then be rotated into an attitude in which they will approach the vertex of the hyperbola tail first (see frontispiece). Shortly before reaching the vertex the motors will be called upon to reduce the ship's speed to that of the capture orbit. However, since the ships are now much lighter than at the beginning of the departure maneuver from the earth orbit, the capture maneuver can be conducted with much less thrust. Only six of the twelve rocket engines will therefore be fired, while the other six will be detached from the ship at the moment of ignition. The remaining six motors, yielding a combined thrust of 132 tons, will fire for 530 seconds. The initial acceleration of a little under 3/10 g will increase during burning to more than 1/2 g. With a weight of 218 tons each, both the round-trip passenger ship and the one-way cargo ship with its landing glider will settle in the circum-Martian orbit.

For the cargo ship the voyage is now over. Nothing is left of it but the four empty elongated tanks which carried the fuel for the capture maneuver, and the six rocket engines, making a total weight of no more than 8.24 tons. The 177-ton glider will now be detached from this exhausted "interplanetary booster" and read-ied for the descent to the surface of Mars (see Plate XXVII).

At departure from the earth orbit, the passenger ship's cylindrical central hull had been surrounded with an array of six tanks for the first and second power ma-neuvers (see Plate XXVI). After jettisoning the four empty spherical tanks, it had performed the capture maneuver near Mars (and any necessary corrections of the circum-Martian orbit) with the propellants in the two extra tanks still attached. For the return flight to earth these two tanks will likewise be removed and the rocket motors will be fed from the 180.4 tons of propellants stored inside the cylin-drical hull itself.

With an initial weight of 237.1 tons the passenger ship, now carrying all twelve members of the expedition, will set out for its return trip to earth. It will burn 129.3 tons of propellants to get back into the circumsolar return ellipse (this figure again includes an ample reserve for necessary corrections). The remainder, 51.1 tons, will be used for the final maneuver, when the speed of the hyperbolic sweep into the gravitational field of the earth must be retarded sufficiently to in-duce the ship into a circular orbit around the earth of 56,000 miles radius (see Fig. 23). Prior to the final maneuver, two more rocket engines will be dropped. All fuel being exhausted, the ship's final weight will be 38.4 tons—2 per cent of its weight at departure. The 98 per cent of the passenger ship's weight used up during the voy-age consisted of propellants, of tanks and rocket engines dropped, and of supplies consumed en route.

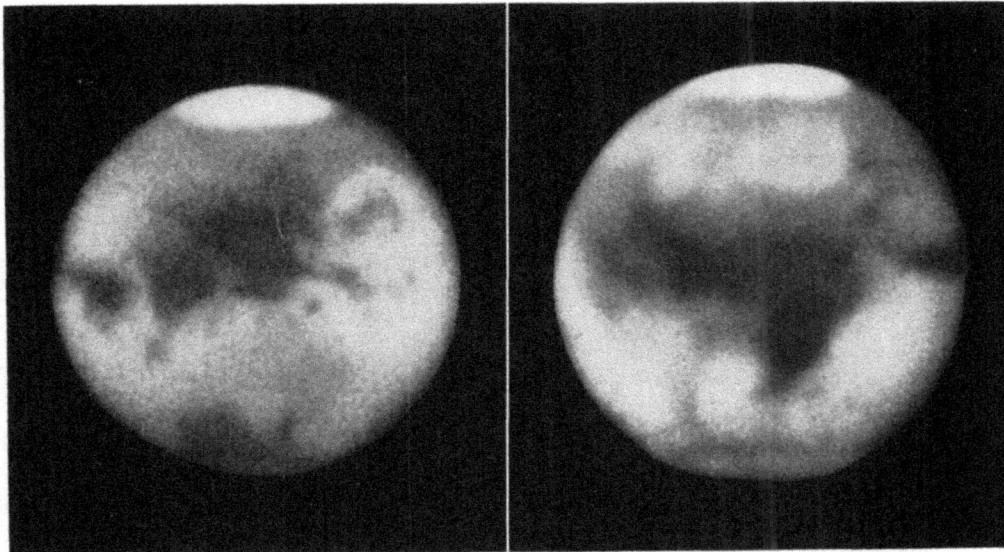

XXV *a*. Mars as photographed by E. C. Slipher at Bloemfontein, South Africa, during the opposition of 1939. The picture at left was taken July 27, the one at right on August 9.

(*Courtesy Yerkes Observatory*)

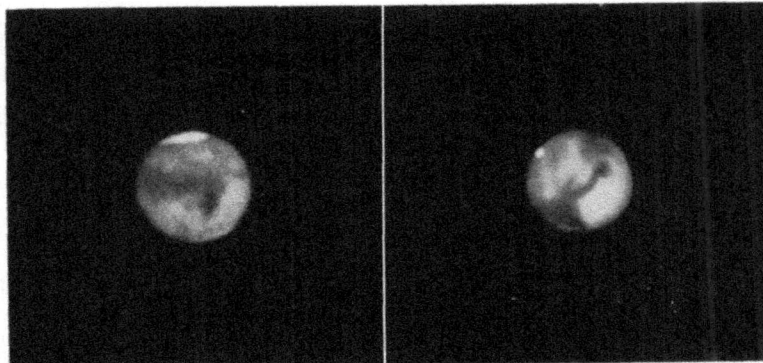

XXV *b*. Photographs of Mars taken at the Lick Observatory, Mount Hamilton, California; on July 20, 1939 (left), and on October 10, 1941 (right). (*Courtesy Lick Observatory*)

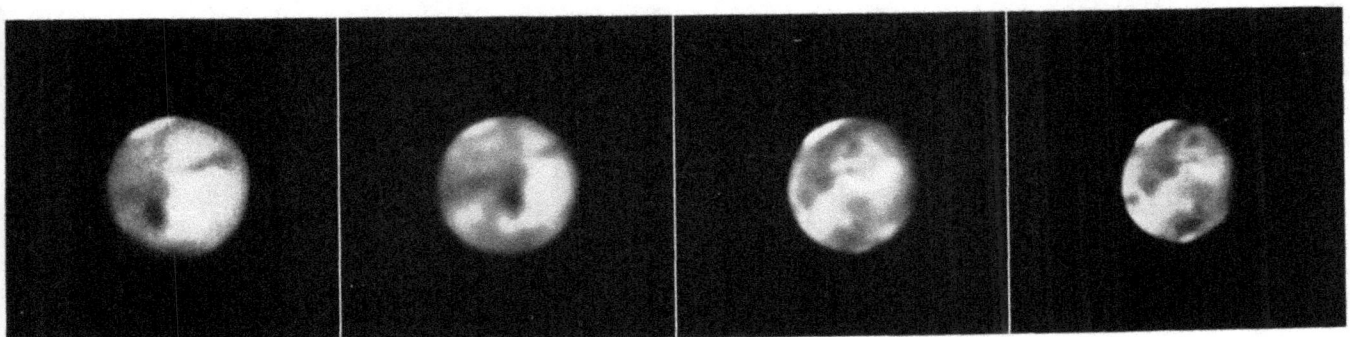

| June 23 | June 26 | July 18 | July 20 |

XXV *c*. Appearance of Mars during the opposition of 1954. (*Courtesy Lick Observatory*)

XXVI. Maneuver 1: The two ships leave for Mars, traveling east in the plane of the ecliptic. They are shown 1330 miles from the earth. Part of the visible portion of earth is the South American continent.

114

XXVII. Having completed Maneuver 2, the ships are orbiting around Mars at a distance of 620 miles from its surface, and preparations are being made for landing.

115

XXVIII-XXIX. Map of Mars showing the currently used names of the main Martian features. Based on a composite map of Mars by Gérard de Vaucouleurs, compiled from photographs and visual observations made during the oppositions of 1939 and 1941 by Earl C. Slipher at Bloemfontein, South Africa, by Lyot, Carmichel, and Gentili at the Pic du Midi, France, and by de Vaucouleurs at Le Houga Observatory, France. (*Courtesy of Macmillan, New York; Faber and Faber, London; Editions Albin Michel, Paris.*)

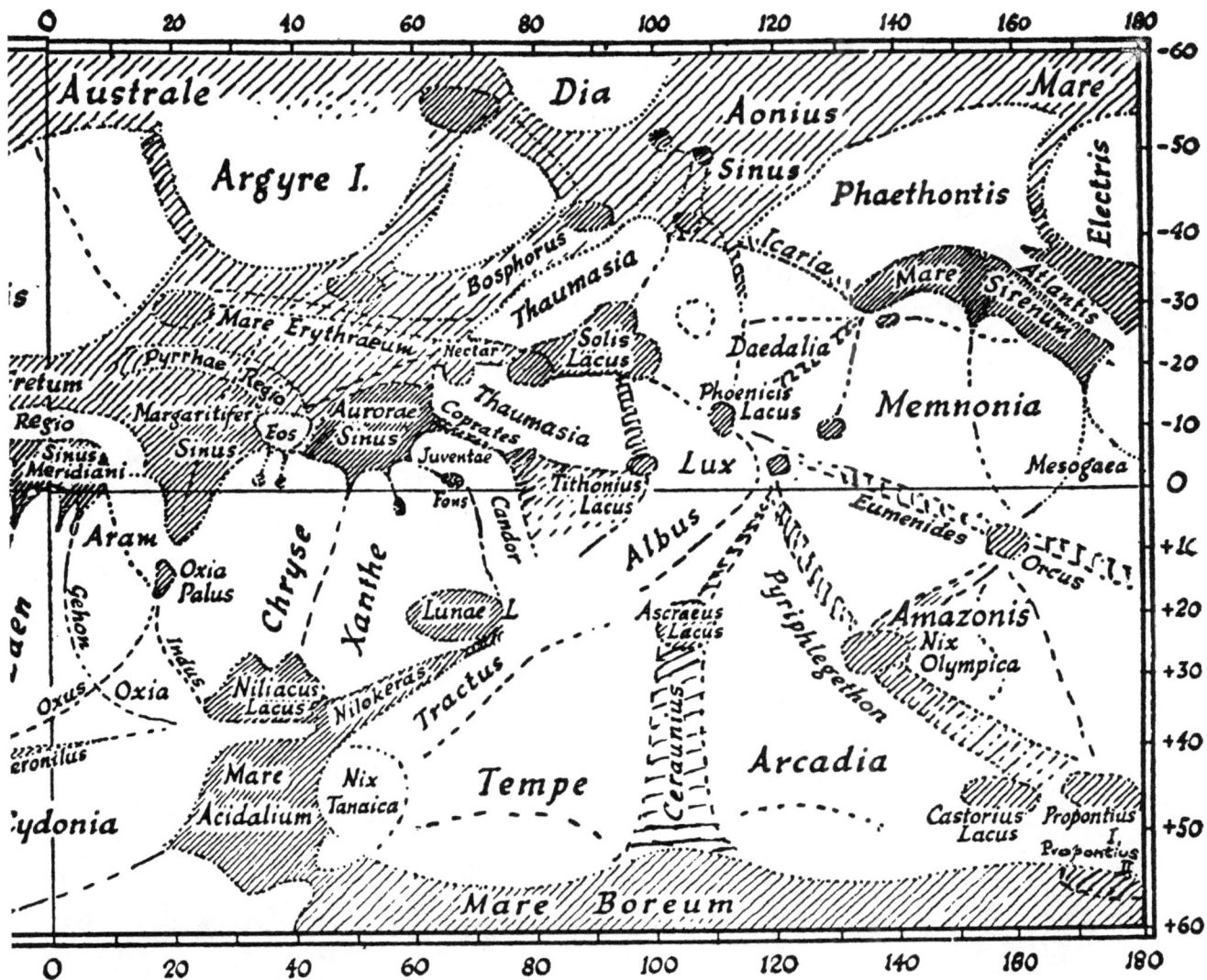

Top scale: 0 20 40 60 80 100 120 140 160 180

−60

Austral e · Dia · Aonius · Mare

−50

Argyre I. · Sinus · Electris

Phaethontis

−40

Bosphorus · Icaria · Atlantis

Thaumasia · Mare · Sirenum

−30

Mare Erythraeum · Nectar · Solis Lacus · Daedalia

−20

Pyrrhae · Regio · Phoenicis Lacus · Memnonia

Margaritifer · Aurorae Sinus · Coprates · Thaumasia

−10

retum · Regio · Sinus · Eos · Juventae · Lux

Sinus Meridiani · Candor · Tithonius Lacus · Mesogaea

0

Fons · Albus · Eumenides · Orcus

Aram · Chryse · Xanthe · +10

Oxia Palus · Lunae L. · Ascraeus Lacus · Amazonis · Orcus

Gehon · Oxia · Indus · Tractus · Pyriphlegethon · Nix Olympica · +20

Oxus · Oxia · Niliacus Lacus · Nilokeras · +30

Tempe · Arcadia · +40

eronius · Mare Acidalium · Nix Tanaica · Ceraunius · Castorius Lacus · Propontius I. +50

ydonia · Propontius II.

Mare Boreum · +60

Bottom scale: 0 20 40 60 80 100 120 140 160 180

117

XXX. The landing craft touches down on a Martian desert.

XXXI. After landing, the wings and undercarriage are disengaged, and the rocket is raised into take-off position.

119

XXXII. The expedition is 270,000 miles from the earth and 30,000 miles beyond the moon. Scorpio is in the background; the bright star just below the moon is Antares.

120

6. OPERATION SPACE LIFT

UNFORTUNATELY Chapter 5 has told only part of the story. So far we have conveniently assumed that the two Mars ships will depart from an orbit 1075 miles above the surface of the earth. But every ounce of their departure weight of twice 1870 tons must first be hauled up to this orbit, where the two ships are to be assembled and fueled and whence they will depart for Mars.

The "space lift" must be carried out by means of a fleet of special surface-to-orbit shuttle rocket ships. In principle, the haul can be made either with a moderate number of flights by very large ships of great payload capacity, or with a larger number of flights by smaller ships of lesser payload capacity. The choice will be based upon a few practical considerations. Larger ships will be more costly to develop, will require larger ground handling facilities, but will be a bit more economical in terms of fuel consumption per pound of payload delivered to the orbit. Smaller ships, necessitating a larger number of flights, call for a more careful coordination of timing between supply and assembly operation in the orbit. They also require that the prefabricated Mars ships be broken down into smaller components, which, of course, complicates the task for the assembly crews in the orbit.

A practical choice may be a three-stage shuttle rocket of 11.0 tons payload capacity for dry cargo. When the ship is carrying propellants for the Mars ships, the payload capacity will be slightly more—11.2 tons—since the propellant payload can be carried in the extra-large tanks of the third stage and no separate payload storage compartment is required. As the two Mars ships, fueled and ready for departure, have a weight of 1870 tons each, approximately 335 supply flights will have to be made for transporting the components of the ships and their propellant supply to the departure orbit. However, additional flights are needed to rotate crews during the assembly of the Mars ships in the orbit and again later on to bring the members of the expedition back to earth from their ship, which will be circling in a rather distant return orbit. The total number of orbital supply flights thus becomes an even 400.

The orbital supply ship is a three-stage rocket approximately 180 feet long and 38.4 feet in diameter at the base (see Plate XXXIV, *right*). Its take-off weight is 1410 tons. The first stage is powered with a battery of rocket engines yielding a total thrust of 2810 tons and operating for 84 seconds. By the time the exhausted first stage is dropped the ship has attained a speed of 1.46 miles per second and an altitude of 24.9 miles, climbing at an angle of approximately 20 degrees with respect to the horizon. The motors of the second stage, yielding a combined thrust of 352 tons, operate for 124 seconds and boost the ship's speed to 3.99 miles per second. With the ship now climbing at an angle of only 2.5 degrees, at an altitude of 39.8 miles, the exhausted second stage is dropped and the nose of the ship forges ahead under the 44-ton thrust of the single-engined third stage. At a speed of 5.13 miles per second, attained at an altitude of 63.3 miles, the fuel supply to the third stage's engine is stopped. In an elliptical flight path the unpowered third stage will now coast halfway around the earth, climbing steadily, until it has attained an altitude of 1075 miles. In a short burst of power, its "apogeal velocity" of 4.114 miles per second is now matched to the exact velocity (4.40 miles per second) of the circular 1075-mile orbit in which the Mars ships are assembled.

The first shuttle rocket to ascend to the orbit has a personnel compartment for fourteen men (including a crew of two) and is equipped with wings which enable it to return to the earth's surface (see Plate XXXIV, *left*). It can carry only 1.1 tons of dry payload in addition to crew and passengers. Thus it is designed primarily for safe transportation of personnel. Before the return flight the burned-out third-stage fuel tanks, together with the third-stage rocket engine, are disconnected in the orbit, and the winged top section, which is in effect a fourth stage, would swing back into the atmosphere with the help of a very small separate rocket engine (see Plate XXXV, *top*). This manned, four-stage version of the supply ship, however, is the exception rather than the rule in the supply operation. The cargo rockets, the workhorses of the Mars-ship assembly operation, will be *unmanned*.

The cargo rocket has no fourth stage, nor can its third stage be returned to earth. Ground crews will launch it at an exactly predetermined instant, and throughout its powered flight it will be controlled by a built-in artificial brain, as a guided missile is. As it finally swings, unpowered, into the orbit, a remote-control radio operator, sitting in the astrodome of the previously launched manned ship, will take charge of its arrival, fire its rocket motor for the speed-matching maneuver, and guide it to match orbit and speed of manned ship.

Each of the unmanned orbital cargo rockets will consume 1226 tons of propellants to carry its 11.0-ton dry payload (or 11.2 tons of propellants) to the orbit. We need a total of 400 supply flights for the entire operation. This figure includes 28 no-cargo flights (one per week) with returnable top stages to rotate assembly crews, and one flight to bring the expedition members to their completed ships. If we plan for an average of two flights every 24 hours, the entire space lift can be completed within approximately seven months. The total propellant consumption for the orbital supply operation will be 490,000 tons, which is a trifle less than the weight of the gasoline used during the Berlin Airlift (see *Aeronautical Engineering Review,* March, 1949, pp. 25 ff.).

The whole space lift for the Mars expedition will be preceded, in turn, by a surface transportation problem. All the ships, all their supplies, and, in short, every item needed for the expedition and its preliminaries must be brought to the ground base from which all space operations are conducted.

The location of this base, unfortunately, is not just a matter of easy accessibility to manufacturing establishments. Even the location of the ground base has to be guided by "astronomical reality," and one important fact demands that it be situated either on the equator or as near to it as geographical conditions permit. A rocket ship on the ground at the equator is, of course, endowed with the velocity of any point of the equator, due to the diurnal rotation of the earth. This velocity amounts to 1520 feet per second, and it comes as a free gift to any rocket ship ascending to an orbit in an easterly direction. Since the orbit in which the Mars ships will have to be assembled is inclined to the plane of the equator the freighter rockets do not receive the full advantage of this circumferential velocity, but even so, take-off from the equator will present a substantial fuel saving, especially when so many flights are involved.

Though this applies to any point along the equator, the choice of the actual ground base is further limited by other considerations. The cheapest way of shipping bulky and heavy freight of any kind has always been by ocean-going vessel, and this holds true for the future too. Consequently the ground base must be accessible to seagoing shipping. Furthermore, there must be a minimum of a thousand miles of open water to the east of the take-off site. The reason for this is that the first and second stages of all orbital supply rocket ships drop back into the

THE EXPLORATION OF MARS

atmosphere and finally strike the earth. The easiest way of preventing them from causing any damage, and at the same time disposing of them, is to have them fall into the open ocean in an area which is not even crossed by regular shipping lanes.[1]

Finally, weather conditions at the launching site should not unduly interfere with the schedule for the orbital ferry operation. While neither an overcast nor even a dense fog would hamper the firings of automatically controlled cargo rockets, severe storms would definitely ground them. Moreover, the base is supposed to handle not only launchings of ferry rockets but also landings of their winged upper stages for personnel transport. While in an emergency these glider stages could land on any airport on earth, they would have to be shipped back to the launching base, because they could not make another flight to the orbit without the help of their first and second stages. It is obvious that reliable and readily predictable weather conditions will be helpful in avoiding awkward interruptions of the supply operation.

All these factors—location on or very near the equator, accessibility to seagoing vessels, more than a thousand miles of open water in a specific direction, and generally good weather—favor the selection of an island in the Pacific Ocean as the most likely place where all these conditions have been met by Nature. This selection will have been made long before the first conference on the space lift for the Mars expediton has ever met, for all these considerations also apply to the operation by means of which the space station will be established. The same base would, of course, be used to supply and maintain the space station.

The island, incidentally, cannot be too small, since it has to have a large harbor, provide space for an airport with long runways, and still have enough area left for test sites, the take-off site (or sites), and a number of other buildings which should be rather far apart. Since the ultimate raw materials for the propellants are nitrogen and hydrogen (air and water), and since large amounts of energy can nowadays be supplied almost anywhere quite economically by means of nuclear power reactors, it might be possible to manufacture the propellants near the ground base to avoid the shipping of the bulkiest and heaviest item over long distances. If

1. It appears to be entirely feasible to recover the exhausted boosters by a combined action of metallic brake parachutes and brake rockets (see *Across the Space Frontier*, pp. 33 ff.). However, the practical value of such a procedure is open to challenge. The construction and operation of ocean-going salvage vessels is costly. The recovered boosters will always require thorough reconditioning and often extensive repairs. Add to this the probability that a certain percentage of the boosters will be beyond repair or even total losses, and it appears quite possible that the costs of salvage might exceed the savings. Since no conclusive answer can be given today, booster recovery has not been considered for the space lift.

not, the transportation of the propellants for the space lift—300,000 tons of nitric acid and 190,000 tons of hydrazine—would correspond to the capacity of forty-one tankers of 12,000 tons displacement.

The choice of an orbit 1075 miles above sea level with a period of revolution of exactly 2 hours as the orbit for assembly and departure may need some explanation here. As has been stated, the entire operation will take seven months if two flights daily are performed. Since the orbit is inclined to the equatorial plane and since the earth revolves inside the orbit from west to east at a rate of one revolution in 24 hours, each crossing of the equator by the "assembly area" will take place a little farther to the west. Since the assembly orbit has a period of revolution of exactly 2 hours, the assembly area, as it crosses from the northern to the southern hemisphere, will pass the equator each time at a meridian 30 degrees farther west. After twelve revolutions, or 24 hours, it will again pass exactly over the meridian of the launching base, which has meanwhile gone through a full circle—12 × 30 = 360 degrees. Only in this relative position between orbit and launching site can the assembly area be reached by the ascending supply rockets. Thus it is evident that a daily flight schedule is feasible only if the orbital period of revolution is an even fraction of the 24-hour day. The 2-hour orbit, which meets this requirement, is low enough for economical supply flights.

Long after the orbital ferry flights have been completed, Operation Space Lift will be called into action once more, to close an important link in the over-all plan. At the end of the interplanetary voyage, the exhausted round-trip Mars ship will not return into the 1075-mile orbit from which it departed 2 years and 239 days earlier. Rather, it will settle in a very high circumterrestrial orbit of 56,000 miles' radius. This is because a substantially larger amount of propellants would be required to induce the returning ship into an orbit as low as the orbit of departure, and, since the return into the earth orbit is the passenger ship's last power maneuver, this extra fuel would have to be carried all the way to Mars and back. It would be ballast throughout all previous maneuvers and thus considerably increase the ship's propellant requirements and initial weight. The over-all propellant bill soars still higher, of course, when we consider that the greater initial weight must first be hauled to the orbit in many additional ground-to-orbit supply flights.

Unfortunately, the three-stage orbital supply rockets are not capable of ascending from the earth's surface direct to the 56,000-mile orbit to pick up the returned Mars explorers. But by use of another simple strategem the costly necessity of developing a special ship for this special mission can be circumvented.

First, one of the winged manned ships will be sent to the 2-hour departure orbit. This first ship will be followed by eight standard unmanned cargo ships, each carrying 11.2 tons of propellants as payload in its extra-large third-stage propellant tanks. (The tanks have a total capacity of 25.2 tons, of which 14 tons are needed for the ascent; hence 11.2 tons of propellants arrive in the orbit as payload.) When all eight "tanker noses" have been guided in by the remote-control operator to settle close together in the 2-hour orbit, they are all refueled to capacity; this requires another ten tanker flights. While this is going on, the airplane-type nose section of the manned ship is systematically deprived of all extra weight. The third stage, which is still attached to it, is disconnected; the wings and aerodynamical control surface and the landing gear are taken off, so that not much is left except the cabin and the navigational equipment.

At this point we have a stripped-down passenger stage and eight fully fueled third stages of the cargo ship, still with their rocket motors attached. They are

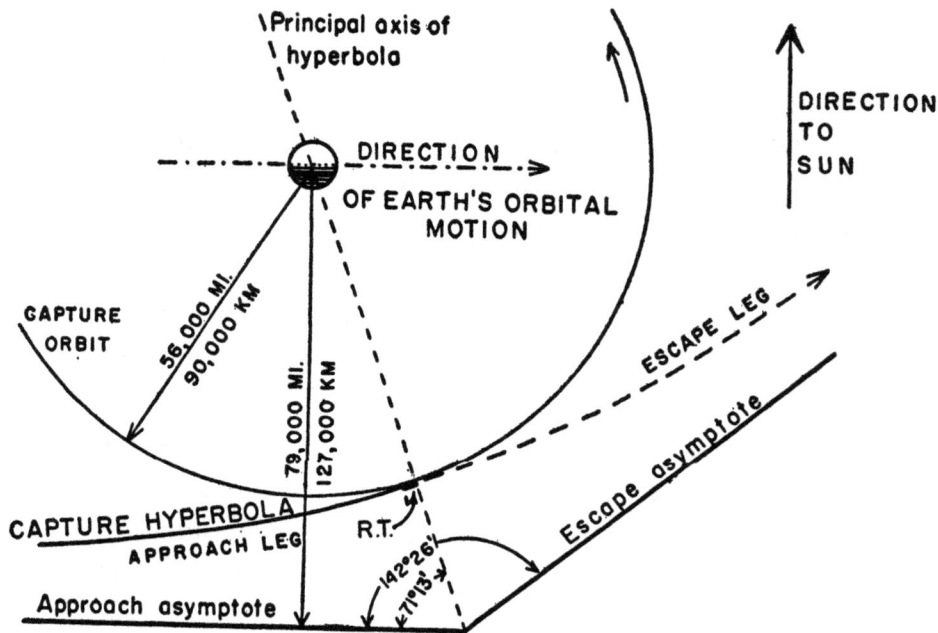

FIG. 23. Maneuver 4—the return into the 56,000-mile orbit around the earth. The short section on the approach leg of the hyperbola, labeled R.T., is the retardation track during which the ship is slowed down. The two engines jettisoned at ignition of Maneuver 4 will continue along the escape leg of the hyperbola.

grouped together with the aid of prepared connecting pieces so that one forms the center with the other seven around it. Then the rocket motors of the seven peripheral stages are taken off and the one of the stage in the center is adapted to receive propellants from all the tanks—a comparatively simple job consisting mainly of rigging two circular connecting propellant lines to the fuel pumps of the only remaining rocket-engine assembly. Then the partly dismantled passenger stage is placed on top of the center of this eight-tank assembly (Plate XXXIX). The result is an awkward-looking but highly efficient vehicle, ideally suited for the relief mission it will have to perform.

After the Mars ship, having returned from its long voyage and established itself in an orbit around the earth and 56,000 miles away, has completed a minimum of one full revolution in this orbit, the precise shape, distance, and position of the orbit will be known. This could be established by optical and radio observations from the ground and from the space station, with cross checks by radar. It can also

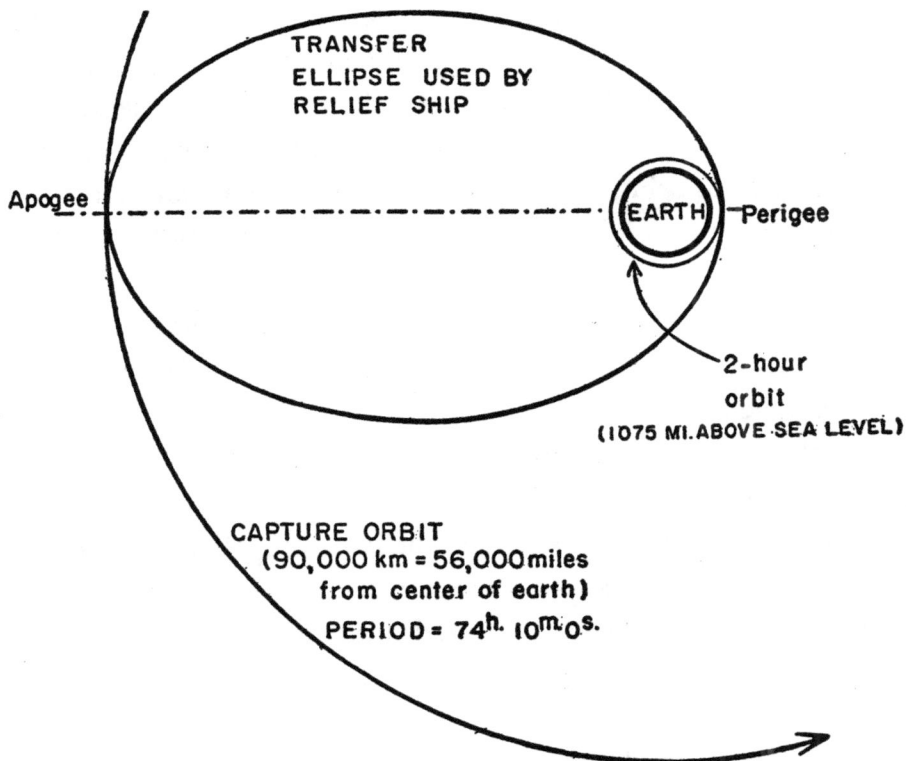

FIG. 24. Transfer operation of the relief ship. A special relief vessel is used to ferry the members of the expedition from their own ship to the space station. The Mars ship stays in its 56,000-mile orbit.

be established by the navigator of the Mars ship. In reality, the various observers are likely to work it out independently and then have their determinations of the orbit compared and checked by a central operations office which will supply definite figures to the relief ship, assembled and ready in the 2-hour orbit. For the first maneuver of the relief operation, the ship will leave the 2-hour orbit 1075 miles above the earth and climb along a half-ellipse to the 56,000-mile orbit of the returned Mars ship. This means that its rocket engine has to produce a velocity change of 1.57 miles per second in addition to the 4.4 miles per second which constitute its orbital speed in the 2-hour orbit. The extra speed will induce it into a long-stretched ellipse which at its apogee (its highest point above the earth) will reach the Mars ship's 56,000-mile return orbit (see Fig. 24). The free-coasting climb through the half-ellipse will last 30 hours. In a short adaptation maneuver, the relief ship will then add 0.78 miles per second to its speed in order to match that of the orbiting Mars ship and take the twelve explorers aboard.

Next, the relief ship will reduce its speed again by 0.78 miles per second so it swings back down to the 2-hour orbit. As it picks up speed during the elliptical fall from the higher to the lower orbit, it will ultimately sweep into the 2-hour orbit with the same excess speed of 1.57 miles per second at which it had departed from it. In the fourth and final power maneuver this velocity difference is reduced to zero.

Back in the 2-hour orbit, the Mars explorers will be transferred to the winged top section of an orbital supply ship for their return to the earth's surface.

Long before the shuttle rockets begin to roar up into the skies to bring their cargoes to the orbit of departure, the two Mars ships will be assembled and checked out in a factory hangar.

A casual glance into this assembly building would convey the impression that many of the design principles used in the deep-space ships have been borrowed from one of the harbingers of the Air Age. The large propellant tanks of thin Fiberglas fabric for the first two power maneuvers will be reminiscent of the helium ballonets, and the filigree tracery of riveted duraluminum trusses in which these tanks would be suspended will resemble the ring frames used in long-departed Zeppelins.

In one corner of the building stands the spherical crew nacelle of the "passenger ship." Blocked upon a flat wooden stand, it looks like a glistening silver balloon some 26 feet in diameter. Within it, a crew of men is working on the interior fittings, having entered by the already completed entrance airlock by which the moderate internal pressure is prevented from escaping. The job in hand is to adapt and fit the electrical wiring and the rubber tubes for the air-circulation system. When completed, these will again be removed like all other fittings, so that the sphere can be collapsed, like the propellant tanks, and in that form freighted up to the departure orbit. The nacelle is re-inflated in the orbit, and its fiber-glass skin covered with thin sheets of duraluminum for protection against meteoric dust.

In the upper part of the sphere is the control deck. This will be the foremost element of the ship after final assembly, being topped only by the antenna mast for ship-to-ship radio and for communication between the orbiting passenger ship and the landing party after it descends to the surface of Mars. The control room is filled with a bewildering assortment of gauges, gyro and navigation gear, electronic equipment, and operating panels.

The two decks beneath the control room contain the living, eating, working, and sleeping quarters, and provide enough space to accommodate all twelve members of the expedition during the 260-day return voyage. There are an electronic food-heating unit, a cold-storage unit, a dishwasher, and also a sick bay and a washroom. Some of the space is occupied by the water- and air-regeneration equipment, and in a secluded corner there is an ingenious installation performing the joint function of an interplanetary garbage ejector and a toilet.

Communication between the various decks takes place through central and concentric openings through which runs a fireman's pole, for the weightless condition prevailing throughout the voyage renders any ladder or stairs superfluous.

At the very bottom of the sphere is the airlock. After the passenger ship is assembled in the orbit, this airlock will lead directly into the large cylindrical hull being assembled in another corner of the large hangar. This hull surrounds the propellant containers for the two return flight maneuvers. Made of duraluminum, it serves the dual purpose of protecting the inner tanks from meteoric grains and of connecting the spherical crew compartment rigidly with the rocket engines in the ship's rear. An array of thermostatically controlled shutters spaced around the hull is used to regulate the impinging heat radiation from the sun in such a manner that the propellants in the inner tanks are prevented from evaporating or freezing.

Bolted to the aft end of the hull is the thrust-carrying framework of the twelve rocket engines which, prior to the assembly of the ship, have been thoroughly checked and adjusted on a static test stand. Each thrust chamber has its own turbine-driven set of centrifugal pumps for the two propellant components, hydrazine and nitric acid. The turbines are driven with the exhaust products from three gas generators—small high-pressure furnaces operated with the main propellants. Two gas generators drive the turbopumps for the eight rigid-mounted thrust chambers in the central core of the rocket-engine battery, six of which will be detached after the first power maneuver and two prior to the last. The third generator supplies the four outer engines, whose thrust chambers are mounted in hinges.

To control the ship's flight path during power maneuvers, these outer thrust chambers are deflected by means of hydraulic actuators which in turn derive their instructions electrically from the gyroscopic guidance system in the ship's nose.

In another hangar is the winged landing craft in which the explorers will glide down to Mars' surface. Compared to the huge wings with a span of 450 feet, the hull looks tiny. And yet only the thinner forward part of the hull will reascend to the circum-Martian orbit after the landing party has accomplished its mission on Mars. The bulkier rear end of the hull is merely a storage bin. It will house the interplanetary radio station, the telescope, the sounding missiles for the exploration of the Martian atmosphere, and the supplies needed for the homebound interplanetary voyage, all of which will be removed prior to the descent to the surface of Mars. In addition, the bin will accommodate all the equipment and supplies the explorers need during their stay on Martian soil.

The large wing area of 24,500 square feet is a concession to the low density of the Martian atmosphere. As the atmospheric density at the surface of Mars is but one-twelfth of that on earth, the obtainable aerodynamic lift of a given wing at the same air speed is likewise only one-twelfth. Fortunately, this drawback is to some extent counterbalanced by the weakness of Martian gravity, which, at the surface of Mars, is only 38 per cent of gravity at the surface of the earth. As a result, the landing weight of the craft of 165 "earth tons" would only be 63 "Mars tons," and this is the weight those wings would have to support when the craft is coming in for the landing on Martian soil. Nevertheless, the weaker gravity on Mars only partly compensates for the greatly reduced atmospheric density. The two figures just quoted—one-twelfth of the available aerodynamic lift versus a weight reduction to 38 per cent—indicate that any given glider, regardless of its absolute size, will have just about double the landing speed on Mars that it would have on earth. Conversely, we may conclude that for a landing on Mars we will have to quadruple the wing area if we propose to maintain the same landing speed as on earth.

7. THE HUMAN ASPECT

THE PUBLIC—or rather the section of the public not associated in any way with modern engineering research—tends to cherish a small number of myths which would evaporate like frozen carbon dioxide on a warm day if they were only subjected to some careful thought and a few practical considerations. The profession of test pilot is a case in point; by most people the test pilot is still thought of as a daredevil who will take a new and untried type of airplane into the sky presumably for the thrill it affords him. The test pilot is thought to be a man trusting to luck—personal luck, pure and simple. Supposing this were so, just which company would let him sit down in the cockpit of a new plane representing several million dollars in development cost?

Contrary to such belief, a test pilot is an especially methodical and careful man. He is a very good pilot with a pronounced sense of responsibility—and usually with the knowledge that a wife and three children are waiting for him to come home. Nor is a test pilot, as it were, born for the task—as a kind of modern equivalent of a musical or mathematical prodigy. Of course he needs some basic aptitude to begin with, but otherwise he is just carefully selected from a great many available men, all of them good pilots by dint of long and intensive training.

All this reasoning, and more of it along the same lines, applies to the men who are going to make the first expedition to Mars. If one of the men on the preliminary lists responds to psychological tests in a manner that indicates that he is the daredevil type, or that he privately thinks of himself as a superman, his name will vanish from the list. If he is of a quarrelsome disposition he will be dropped from the expedition—though not necessarily from the whole project, if he is otherwise outstanding—as surely as if a physical check-up had revealed cirrhosis of the liver.

The men selected for training (which does not mean only those who will actually go, since some may still turn out to be unfit, and a reserve force is needed for possible last-minute replacements) will, of necessity, have many traits in com-

131

mon. They will have to be physically sound, of course; and a few physical types, such as very large and excessively tall men, will almost automatically be excluded. They will probably be at least in their late twenties, simply because they will have reached that age in the course of acquiring the necessary experience and all the knowledge they need. They will be men who are quietly competent, with an outstanding capacity to learn, an exceptional ability of adaptation, and a preference for working in and as a team. They must have a sense of humor and combine a practical outlook with unlimited imagination. The neighbors' children may think they are dull, in spite of their glamorous jobs. For they will be picked from among the men to whom space is something familiar.

By the time the Mars expedition is in its preliminary planning stage, space stations will have been an accomplished fact for years, and therefore there will be many—possibly as many as a thousand—men who will have spent some time on space-station duty. There will be at least a hundred rocket-ship pilots who have flown supply ships to the space station; there will be many more who have flown in such supply ships in a capacity other than pilot. All of them will have experienced repeatedly, and some of them many times, how it feels to go through the cycle of repeated high acceleration in the take-off and powered flight of a three-stage ship. To them such terms as "coasting up to an orbit," "adaptation maneuver," "retardation maneuver," and "re-entry into the atmosphere" will not be just words of which they know the precise meaning, but will represent living reality. These men, the men with space experience, form the pool from which the preliminary list can be made, or rather those whose applications are likely to get them on the list. Because of their background they will know a great deal of what the expedition members have to know. They won't need astronomy lessons. They all will have a good understanding of celestial mechanics and a good working knowledge (at the very least) of rocket ships and rocket propulsion. They will have passed the physical check-ups and the psychological tests. In short, they are a group which has already been selected for a very similar task. All that needs to be done is to make a finer, or more stringent, selection and to provide them with special training.

Since this is going to be an expedition where twelve men will be on their own for more than two years, without any possibility of outside aid except such instructions as might be transmitted by radio, versatility is a condition and multiple training a necessity. The radio operator may fall sick and be unable to perform his duties for weeks—there is no way of predicting the exact state of health of any

132

individual for more than two years in advance. The chief engineer may suffer a nervous breakdown as a result of the heavy load of responsibilities he had to endure during the months and months of drifting through the lonely vastness of interplanetary space. The ship's doctor may die,[1] or the navigator be killed or seriously injured in some manner. Logically, then, the radioman must be able to take the place of the navigator, the co-pilot of the glider the place of the chief engineer, while at least three men of the crew should have a fair amount of training in medicine and simple dentistry. In this respect the schooling for the expedition differs from the training for space-station operations, where not only is the crew larger but ground base is not much more than two hours' flight away.

In addition to classroom instruction the expedition members will have to undergo much actual training, most of it in simulating devices which serve the dual purpose of familiarizing the man with his duties and showing how he reacts. One of the devices that will be built and used is the "control-deck simulator." In appearance it will be a dome-shaped room which is an exact replica of the top deck of the spherical crew nacelle of the passenger ship.

For a simulated departure maneuver from the circum-Martian orbit, for example, captain, navigator, radioman, and engineer will be strapped to their contour chairs, surrounded by a maze of instrument panels and radio and gyroscopic gear. Through the intercom system, they exchange tense messages, unintelligible to anyone not graduated in Advanced Spaceman's Latin. A subdued thunder emanating from a loudspeaker indicates that the rocket engines are firing. (Airless space cannot propagate sound, of course, but the engine noise would still be heard, since it is carried forward to the crew compartment through the ship's structure.) A battery of manometers in the engineer's huge instrument panel indicates the combustion-chamber pressures in each of the rocket engines. Next to it, a "mixture-ratio indicator" shows whether the flow rates of hydrazine and nitric acid are correct. There are remote indicators for tank pressures, revolutions of the turbopumps, cabin pressure and temperature, deflections of the hinge-mounted control motors, and a host of other important data. Beneath the instrument board is a console, looking like a scaled-down version of a switch-position indicator in a railroad yard, which indicates by means of dark and bright lines and green and red lamps which valves are open and which are closed.

1. During the German oceanographic expedition of the *Valdivia* in 1898–99 the crew of the fairly large steamer and the whole scientific staff were in the best of health and spirits; the only man lost during the whole expedition was the ship's doctor.

THE EXPLORATION OF MARS

The eyes of the navigator are fixed on a television screen on which an image of the ship's contours indicates the latter's attitude in space with respect to the gyroscopically stabilized platform. (This platform serves as a reference system in outer space, where such notions as "up" and "down" become meaningless.) The same screen also indicates the ship's displacement from the prescribed flight path, which is represented by two crossing light beams. With one glance the navigator can thus see whether the ship is right or left of the path, above or below it, and whether the nose is pointing in the right direction, in order to correct the deviation. Beneath the screen is a long scale with two needles traveling from left to right along its length. The lower needle indicates the expected velocity build-up under the steady push of the rocket engines, whereas the upper one shows the actual velocity. If the upper needle trails the lower one, the ship is moving too slowly; if it runs ahead, the speed is increasing faster than it should.

Suddenly, the tense exchanges become downright nervous. We hear the engineer yelling something into the intercom, and the captain yelling back. One of the lights on the engineer's panel is flickering a red warning as he fumbles excitedly for the correction switch. Eventually, the light goes green again and a relaxed smile returns to his face.

The crew knows that "The Devil" had played them one of his tricks. For outside of the control-deck mock-up there is a panel which would not be in a real ship. Behind it sits a man with a permanent grin. Through his intercom he can hear distinctly the piteous conversation between engineer and captain as he produces readings calculated to drive them to desperation. Hearing their planned corrective measures, he can block their success by a turn of the wrist. There are all kinds of malfunctions he can conjure up to plague the crew. Suddenly, in the midst of a power maneuver, the respiration blower might stop. Or one of the cables leading to the steering gear might show a suspiciously high amperage. The worst and most emotionally disturbing trick he can play is to indicate on several instruments simultaneously a malfunctioning of the rocket engines. This would include a most lifelike imitation of stuttering or howling of the otherwise steady growl of the exhaust. Then in a matter of seconds the captain and his team have to do the right thing or else endanger the entire expedition.

Of course, the control-deck simulator has no real danger attached to its operation. An error and its consequences can be thoroughly discussed afterward, and the best means of preventing or correcting thought out at leisure. Any specific "sequence of events" can be repeated and repeated until the correct response has been drilled into the crew to the point at which it becomes a conditioned reflex.

The control-deck simulator is only one of the many training devices set up to prepare the crews of the Mars ships for their great adventure. In another room there is a complete mock-up of the electrical system of the passenger ship, especially rigged up for trouble-shooter training. An instructor might tear a single wire out of one of the hundreds of plugs. Stopwatch in hand, he then would observe how his trainee goes about locating and correcting the trouble. There is a complete working mock-up of the air- and water-regeneration plant, where maintenance jobs such as cleaning filters, changing gaskets, and replacing valves can be practiced, and emergency situations, such as "no juice" effects or blower trouble, can be reproduced in all their actual horror.

A synthetic training device for the flight crew of the winged landing craft consists of a mock-up of the pilot's cockpit complete with all instruments. In it, the flight crew can undergo the whole gamut of the landing procedure on Mars from the departure out of the orbit to touchdown. An elaborate electronic flight-path simulator accurately reflects onto the instrument panel the craft's response to control movements. Even the Martian landscape, rolling along beneath, is portrayed on the colored television screens which take the place of the canopy windows.

After such training on a simulating device all the expedition members who might be called upon to perform the landing on Mars may do some actual practice landings with a glider that has been designed to have the same handling and flying characteristics in earth's atmosphere as the landing boat will have in the Martian atmosphere. It will have a landing speed which is most unusually high for a glider, and the pilots can practice setting it down, first on firm ground like that of Muroc Dry Lake in California, then on salt flats, and finally in an Arizona desert area.

The most surprising synthetic trainer, however, is the navigator's training device. It consists of a black hollow sphere, approximately 60 feet in diameter. Gimbal-suspended and gyro-stabilized in its center is a replica of the passenger ship's astrodome, with a seat for the trainee. The great hollow sphere surrounding him is covered with thousands of tiny holes spaced in the pattern of the stars in the heavens. The hollow sphere is surrounded by a second outer shell. The light from powerful electric bulbs located between inner and outer sphere shines through the holes and makes them appear to the trainee like the artificial stars in a planetarium. Sun, earth, moon, and planets can be projected in any desired size upon the dome.

THE EXPLORATION OF MARS

The purpose of this impressive installation is to train the expedition's navigators in the complicated task of determining any deviations of the Mars ships from their predicted flight path and setting up the necessary corrective maneuvers. During the months of unpowered flight through interplanetary space the navigator's task would be relatively simple. With an instrument resembling a sextant he would measure once a day the angles between the sun, the moon, or any of the closer planets, and some fixed stars nearby, and thus determine the ship's position in space. But his job would be far more difficult, and time much scarcer, when one of the two tricky capture maneuvers near Mars or near earth was imminent, as, for example, the maneuver of return into the 56,000-mile circumterrestrial orbit.

The great shape of the earth's sphere, half illuminated by the sun, appears on the velvet-black backdrop of the planetarium's dome. As the ship gradually draws nearer, the great multicolored shape covers and uncovers various fixed stars as it moves along, seeming to grow in size to the eyes of the anxious navigator as he lies in his astrodome. He measures the angle between the earth's center and certain nearby stars and punches his readings into a keyboard in front of him. He punches another set of keys at the instant that certain of the stars are obscured by the earth's rim. As a check, he makes a parallax measurement on the moon and punches again.

The keyboard conveys the navigator's readings to an elaborate computer which, within a matter of minutes, will figure out the extent to which the actual hyperbolic fall toward the earth deviates from the prescribed path. If the deviation is not great enough to warrant a further corrective power maneuver, the impending retardation maneuver will be set up to match the actual approach path. The procedure then becomes almost automatic. Should the computer determine that the ship is coming in on, say, Hyperbola no. 237, it will select magnetic guidance tape no. 237 for the impending power maneuver. Guidance tape no. 237, along with hundreds of others, has been prepared long before the expedition departs and is stored in a dispenser operating on the principle displayed by the common jukebox. Once inserted into the ship's guidance mechanism, the tape will do the rest of the job automatically. First, it will rotate the ship into the attitude which it must have in order to move tail first at the time it passes through the vertex of the hyperbola. Second, it will set the timing device to fire the rocket engines at the precise predicted moment. Third, it will set the integrating accelerometer. This is a device that measures the change of velocity (not the velocity itself) that takes place as a result of the firing of the rocket engines, and when the change has added up to the proper figure it actuates the relays that cut off the fuel flow to the engines. After this retardation maneuver the ship should be in the circular orbit around the earth that is a part of the over-all plan.

This kind of training will go on and on until everybody, including the crew members themselves, is fully satisfied that no additional pertinent instruction is possible. Meanwhile, of course, the supply ships have been built and their crews have been trained, and the Mars ships also have been built and undergone all the tests that are possible in the factory.

Then the Space Lift begins.

After their preliminary assembly and checking in the factory hangar the Mars ships have been disassembled. Along with some of the collapsible propellant tanks, the Fiberglas hull of the spherical crew nacelle of the passenger ship has been deflated and folded into the storage hold of one of the unmanned cargo ferry ships. The sectional frames of the duraluminum support structures for the outer tanks have been stowed away in other cargo holds. The large wings of the landing craft have been taken apart into 6-foot sections and all screw joints carefully numbered and coded to facilitate the assembly job in the orbit. The landing craft's cavernous cargo bin, as well as the passenger ship's central hull, have been broken down into slices which can be quickly and easily put together in the orbit by means of connectors and latches. For the forward portion of the landing craft's hull, however—that part of the vehicle which will finally reascend from Mars to the circum-Martian orbit, the plans of the Space Lift provide a noteworthy exception. Being a full-fledged rocket complete with engine and propellant tanks, it will be placed on top of a freighter rocket's two booster stages, in lieu of a standard third stage, and flown to the orbit under its own power.

The stowage of the delicate components being an intricate task, the storage holds of the cargo rockets are loaded in the factory itself. As the bulk of the loads varies to a great extent, the cargo noses are different in length, but all have the same rear diameter of 7 feet to fit them onto the third-stage tank sections of the freighter rockets. Of course, in order fully to utilize their payload capacity, each nose is loaded with a full 11.0 tons of cargo.

In a harbor near the factory the cumbersome crates containing the cargo noses are lowered into the belly of a seagoing freighter which takes the precious load to the launching base, in the Pacific Ocean.

The launching plan for the Space Lift is synchronized exactly with the assem-

137

bly operation in the orbit. Accordingly, after the first manned "guide ship" has reached the assembly orbit, an unmanned cargo flight will deliver the collapsed crew nacelle for the passenger ship. The next cargo flights will bring liquefied air to inflate the nacelle and all equipment needed to sustain life permanently within it. This nacelle will serve as temporary living quarters for the assembly crew.[2]

Further cargo flights will deliver a number of propellant tanks for the Mars ships. This permits the propellant-loading operation to begin without delay. Working in space suits, the assembly crews will rig connecting hoses to the arriving tanker rockets and transfer the hydrazine and nitric acid into the receptacles of the as yet incomplete Mars ships. Meanwhile, the assembly work itself will continue. Two cargo rockets will be launched at a few minutes' interval every 24 hours, the steady stream of "tankers" being only occasionally interspersed with "dry cargo" and "personnel" rockets.

One of the last cargo flights to arrive in the orbit will be the nose section of the landing craft, which performs the last portion of the climb under its own power. After the assembly job is finished and the complete landing craft connected with its "interplanetary booster," a lengthy inspection period will begin. For several weeks every component and every instrument will be checked for proper function, setting, and calibration. After the component tests, over-all systems tests will be made.

At last the day will come when the chief inspector pronounces the two Mars ships "in all respects ready for space." The last supply flight will bring the members of the expedition themselves.

2. It may be possible to provide more comfortable living quarters for the assembly crew if the Mars ships can be assembled in the proximity of a space station. It is likely that more than one space station may be in existence by then, but they will all have been built for other purposes and may have to move in orbits different from the orbit in which the Mars ships must be assembled.

8. EXPEDITION TO MARS

WHEN the day of the departure of the expedition to Mars dawns on earth the newspapers will tell their readers that the personnel rocket carrying the members of the expedition took off the day before. As the time for the departure from the orbit—established to the split second by computing machines—comes near, the men strap themselves into their contour chairs and go once more through the routine checks. The procedures are all thoroughly familiar, the men are all used to space, and nothing, except their expectations, is really new. But because of these expectations every one of the explorers thinks he can hear his heartbeat drumming through the confused hum of the inverters and the whine of the gyroscopes.

For now the time has come—this is it.

The last remaining minutes and seconds are being counted down. The countdown comes from loudspeakers in the two ships. It comes from loudspeakers in the space station. It is followed in the various observation stations and it is heard in countless homes where the television screens show two apparently small ships against a background of black immensity.

At X minus 4 seconds a deep rumble goes through the two ships—the rocket motors are burning at "ignition stage." Under a slight helium pressure, applied against the flexible displacement bags in the starter tanks,[1] the two propellants are being fed at a comparatively slow rate into the twelve thrust chambers, where they ignite spontaneously. The ships hardly move, but this preliminary stage is necessary before the "main stage" can be thrown on, just to make certain that there is a "pilot flame" in each rocket chamber. At X minus 1 the ignition stage is well established,

1. Since it is uncertain whether the fuel pumps can succeed in establishing a proper fuel flow when the ships are under weightless condition a small amount of acceleration has to be produced first. This is accomplished by having separate tanks (which can be pressurized by means of a compressed gas) discharge fuel into the rocket motors to produce this initial acceleration. The bags in the starter tanks merely serve the purpose of separating the liquid propellants from the pressurizing gas under weightless condition.

139

THE EXPLORATION OF MARS

and at the zero moment the main stage goes on. The turbopumps run at full speed, and with a pressure of several hundred pounds per square inch the propellants are injected into the combustion chambers. The rumble becomes a thunderous roar; the thrust very rapidly builds up to its full value of 396 tons. Ponderously the two large Mars ships begin to move visibly. The thunder and the reverberations of the rocket engines last for a little over 15 minutes. Then, as suddenly as it began, the roar subsides. The pitch of the whining gyros declines, and soon only the rustle of the ventilation blowers remains.

The 260-day coasting flight to Mars has begun.

To the members of the expedition the terms day and night are now meaningless words, or words which have a meaning only via memory. For they are now in space, where the sun always shines, where life is regulated by ship's time, and ship's time is established by authority. The eight men who travel in the passenger ship will close all port covers promptly at 2000 hours, expedition time. The four men who travel in the landing craft's pilot compartment from which the cargo ship is controlled will do the same. As punctually at 0700 hours "next morning" the port covers will be removed again. Time for the individual aboard is now governed by the schedule of watches, for even during free-coasting flight watch duty aboard the Mars ships calls for far more activity than seems likely at first glance.

One important duty is checking the temperature of the propellants. The temperatures in the tanks are kept constant by thermostats operating radiation shutters similar to Venetian blinds. Since the angular attitude of the ships need not be controlled during unpowered flight, both ships tumble slowly as they coast along their elliptical path. As a result, some of the outer tanks may temporarily be shaded by the others. If such a condition prevailed for an extended period of time, the liquids in the tanks not reached by any sunlight might cool too much or even freeze. Fortunately, the heat capacity of the outer tanks is so large that even in this case the temperature would drop very slowly. Nevertheless, the watch-keeper will have to observe the tank temperatures and, if necessary, activate the attitude-control flywheels to rotate the ship out of such a "prohibited attitude."

Then there is the air-conditioning system. The crew spaces in the Mars ships are pressurized with an atmosphere very different from that on earth. At sea level, the terrestrial atmosphere has a pressure of 14.5 pounds per square inch and consists of 21 per cent oxygen, 78 per cent nitrogen, and about 1 per cent of other gases. In order to save weight, the pressure in the living spaces of the Mars ships has been

140

reduced to 8 pounds per square inch. To compensate for the reduced pressure, the oxygen content has been increased to 40 per cent, and the nitrogen has been replaced by helium, which combines the advantages of less weight and greatly reduced danger of air embolism in case of a sudden accidental drop in pressure.

Temperatures, humidities, pressures, and oxygen content in the living spaces are automatically controlled, but the watch-keeper has to make regular tests. One of the hazards of extended living in an artificial atmosphere so completely isolated as that in a space ship is the ever-present danger of accumulative poisoning. Toxicologists have established that in the routine of an average household no less than twenty-nine different poisons are produced. For instance, in the rather prosaic process of frying an egg a very potent poison called acrolein may be formed. Of course, in a home this does not constitute more than a nuisance, since the coughing spell resulting from the burned egg white will cause the housewife to open a window or turn on a fan. In the closed air circulation system of a space ship, however, the same incident could easily have serious consequences, as the poison, unless properly filtered out, would be recirculated and rebreathed over and over.

Most toxic compounds can be removed from the air cycle by supercooling the air (thus freezing out the offensive ingredients) or with the help of active carbon and other chemical filters. Nevertheless, the complicated technical equipment installed in the crew compartment of a space ship constitutes a continuous source of possible contamination of the artificial atmosphere. An instrument may break, admitting some mercury to the air cycle. A contact may burn, with resulting evaporation of a tiny amount of some other toxic material. Or an electrical short circuit may cause some cable compound to scorch. For this reason the watch-keepers have to run periodic analyses on the purity of the recirculated air.

Should such tests reveal any traces of poison, the man on duty will first attempt to remove it by inserting selected filters into the air-circulation ducts. In case this does not remedy the situation, his last resort is slowly to vent the air in the living quarters out into the surrounding vacuum, while simultaneously replacing the losses from the storage containers of liquid oxygen and liquid helium.

The temperature in these containers, as well as the temperatures in the water tanks and food compartments—all of which are quite different—have to be read and logged at regular intervals, to make sure that none of the supplies needed to sustain life in the loneliness of interplanetary space is lost or spoiled through oversight or a malfunction of a thermostatic control mechanism.

The electrical power supply system requires the most constant supervision. It is the heart of the whole complicated system of annunciators, remote-reading gauges, and such, which keeps the man on watch informed on the condition of the entire ship. Electricity also feeds the motors of the attitude-control flywheels and turns the temperature-controlling blinds to their appropriate angles. Above all, it whirls the pumps and blowers which maintain the air-conditioning system with all its intricate controls for purification, temperature, oxygen content, pressure, and humidity. Should the small nuclear-driven turbogenerator halt, it would be but a few hours before the battery's exhaustion would bring air circulation to a stop.

In case of a current-supply failure, the watch-keeper would sound an alarm and awaken those crew members who are off duty. While the chief electrician attempted to fix the trouble, the standby solar battery would supply the ship with a bare minimum of emergency power. In case of a major breakdown the sister ship would be requested to maneuver itself into a position just a few hundred feet away. A crew member would don a space suit and bring an emergency cable over to its external power outlet. Through this umbilical cord the vital electricity would be fed to the stricken ship until the repairs were completed.

Five days out. The distance between the Mars ships and the earth has become almost a million miles, and the earth now appears about the size that the moon seems from earth. Since only the right half is illuminated by the sun, it looks like a waxing half moon. But that half shines so brightly that no contrasts upon it can be distinguished with the naked eye.

Some distance from the earth a luminous disklet of about one-quarter earth's diameter is visible—the moon, following the earth on its path around the sun.

The retarding pull of the earth's gravitational field has gradually diminished to virtual zero, as the two ships, flying in the direction of the earth's orbital motion around the sun, enter the long elliptical path to the Martian orbit.

As a result of slight errors in the automatic guidance systems the two ships may have drifted some fifty miles apart during the first five days of the journey. Reflected in the bright sunlight, the cargo ship would still be visible as a conspicuously brilliant star, but the desire to be able to render mutual assistance in case of an emergency calls for a closing of the "formation." With the aid of radio tracking (performed by a number of unmanned, automatic receiver stations spaced along the departure orbit) and by careful timing of star occultations behind the rims of the earth and the moon, the navigators establish the extent to which each ship has strayed from the prescribed path. A short radio exchange between the two ships confirms what both navigators have found independently: the cargo ship's track is quite satisfactory but the passenger ship is lagging a bit behind. A split-second burst from its four deflectable rocket motors increases the velocity of the passenger ship by approximately 1 foot per second. Three days later it has caught up with the cargo ship. Its four steering motors are now rotated through 180 degrees and fired again for a fraction of a second to match speeds. The formation will then keep together until new star fixes taken by the navigators indicate that the circumsolar flight path must be corrected again.

Seventy-three days out.

Everybody knew from the flight-path prediction that this day would present a spectacle never before seen by human eyes—a "transit" of the earth and the moon. At a predicted time the earth and its satellite will pass across the flaming surface of the sun. To the navigators this is also a unique opportunity for making a particularly precise check on the position of the expedition.

At first, the ships' elliptical path kept advancing them away from the earth at an excess speed of 1.88 miles per second. Then the sun's gravitation began to retard the "uphill" swing. The speed had day by day diminished until the ships' angular velocity around the sun had been reduced to less than that of the earth. The earth had begun to overhaul the ships on its inner circumsolar orbit. Seventy-three days after departure, it will pass between the ships and sun at a distance of some 12 million miles (see Fig. 20, *left*).

During the previous few days the bright double star made up by the earth and the moon had become dimmer and dimmer until it finally disappeared completely. Earth and moon are turning their night sides to the receding Mars ships.

The crew crowds the portholes, dark glasses before their eyes. At last they see a tiny black spot across the flaming corona of the sun. Very deliberately it moves from one side into the full glare of the fiery ball. An hour or so later, another spot appears, even smaller than the first, following toward the sun's center.

The total transit lasts for about eight hours. The navigators take two series of measurements. One is concerned with the angles, measured against adjacent fixed stars, under which the earth and the moon are traveling across the sun's disk. This gives them accurate information on the deviation of the voyaging ellipse from

143

the plane of ecliptic. The other set of measurements involves the exact times of the transit—that is, the accurate instants of the first and last contacts of the two little black disks with the rim of the sun. When compared with "ship's time," provided by an ultra-accurate, crystal-stabilized chronometer, these instants furnish exact information on whether the ships are running behind or ahead of schedule on their elliptical path around the sun. Both measurements combined will indicate how any corrective thrust maneuver, if necessary, must be laid out in order for the expedition to meet Mars at the appointed rendezvous point—still 187 days away.

After two more months have passed, weariness and extended inactivity make themselves felt among the crew. Personalities are beginning to wear on one another with resulting tensions. A limited exchange of personnel between the two vessels is undertaken to alleviate the unbearable monotony.

One of the few cherished relaxations is the daily radio broadcast from the earth, beamed at the receding Mars ships by a special booster transmitter near one of the space stations orbiting around the earth. Despite the 40 million miles which by now separate the ships from the home planet, reception is as clear and undistorted as though it came from a local station around the corner. The broadcasts usually consist of news, a lecture or so, and music, although from time to time some regular program is relayed to the ship.

Some of the crew members one day don space suits and leave the crew nacelle of the passenger ship through the airlock, and, after a short burst from the reaction pistol, coast over to the cargo ship, which hangs, apparently motionless, about a mile away. Their objective is to see the two-way radio station at work. Visible as they approach the cargo ship is a duraluminum mast which protrudes from the landing craft's cargo bin and carries at its end a swivel-mounted, parabolic dish, in every respect similar to those used in van-mounted radars. This is the combined receiving and transmitting antenna of the interplanetary radio station. The parabolic dish, with the actual antenna rod in its focus, beams the outgoing radio waves in the direction of the earth like the reflector of a searchlight. When receiving, it concentrates the intercepted waves upon the antenna rod, as a mirror can concentrate sun rays to burn paper.

Inside the radio room, which is installed in a separate pressurized container imbedded in a well in the large cargo bin, the visitors learn that the gain incurred from this beaming and focusing of the transmitted and received energy is only part

XXXIII. A dust storm on Mars; if seen from earth, it would be listed as a "yellow cloud."

XXXIV. Orbital rocket ships. At left, a four-stage passenger ship, with an airplane-like passenger section which can return to earth; at right, a three-stage cargo ship. The ships are alike except for the addition of the fourth stage on the passenger ship and the difference in the tail fins.

(*Courtesy Walt Disney Productions*)

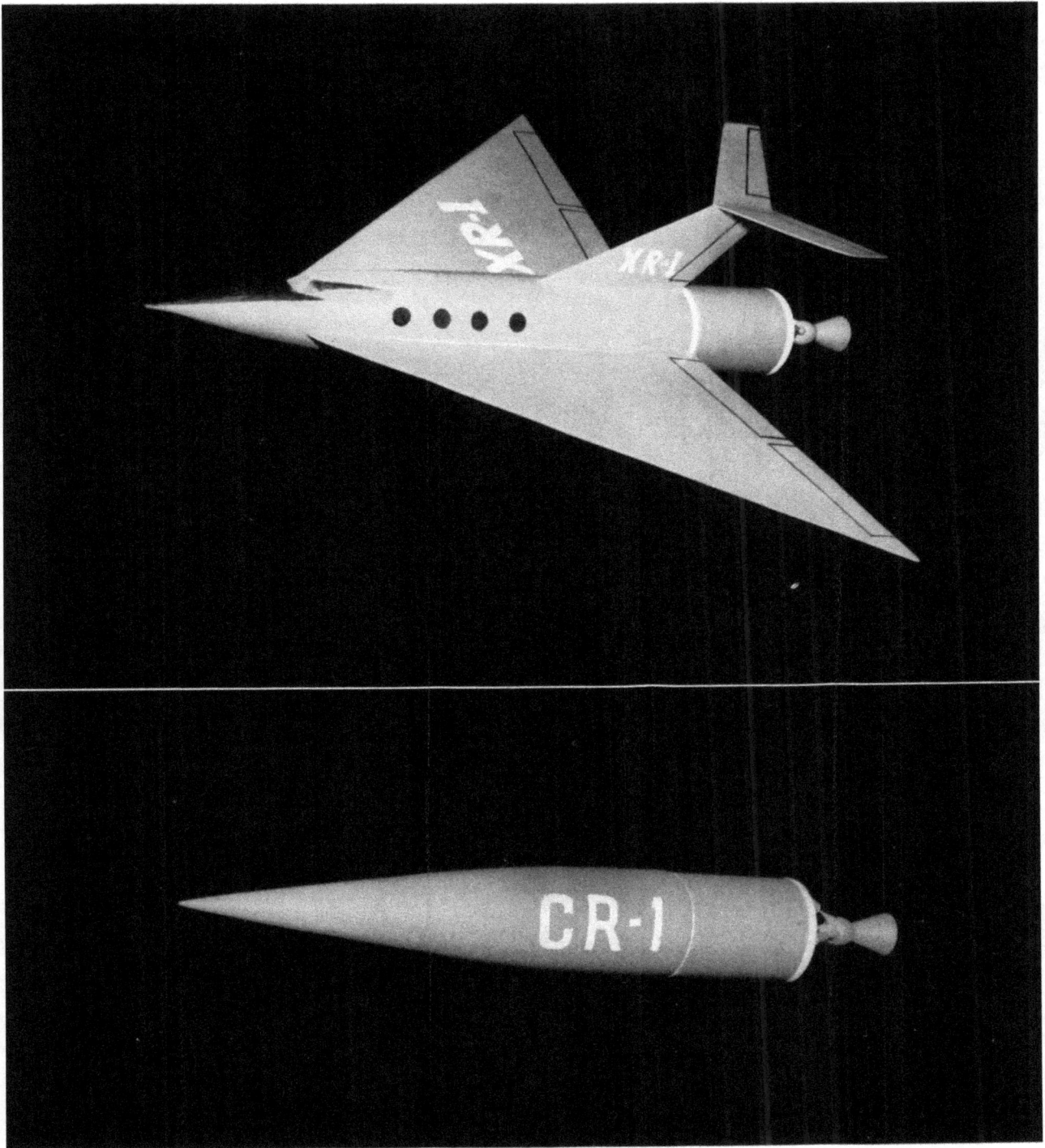

XXXV. (*Above*): The earth-returnable fourth stage of a passenger ship, with the third stage still attached. The third stage will be detached before the return flight. (*Below*): The third stage of a cargo ship, which remains in the orbit. This stage will be made in different lengths, depending on whether the cargo to be carried is dry freight or liquid fuel.

(*Courtesy Walt Disney Productions*)

XXXVI-XXXVII. The ground station has been set up on Mars, and the explorers are beginning to investigate the vicinity of the landing site.

Chesley Bonestell

149

XXXVIII. With the ship 180,000 miles from home, earth is again a planet and no longer a bright star in the sky. The ship will soon enter the 56,000-mile orbit around the earth.

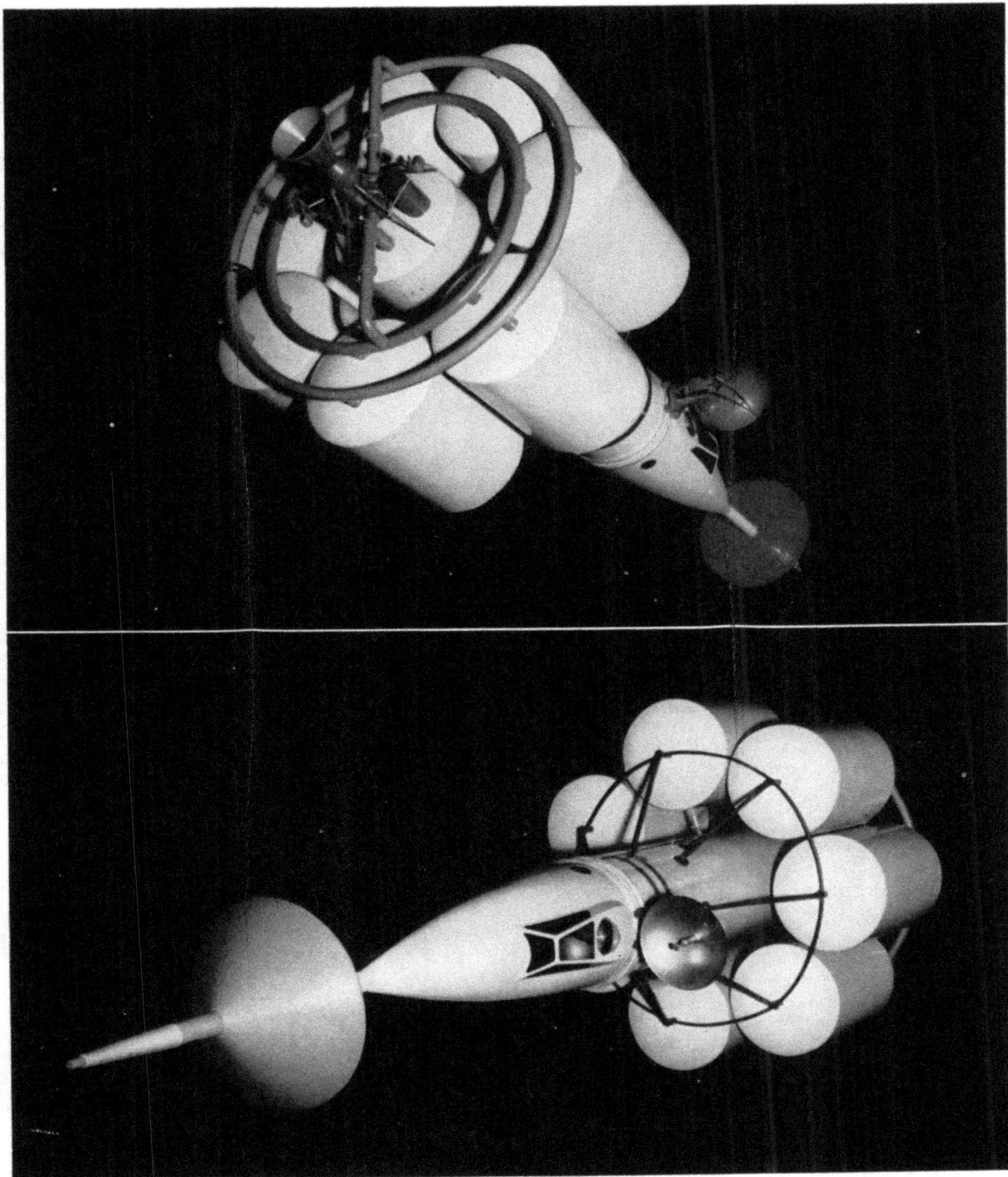

XXXIX. Front and rear views of the relief ship. The relief ship consists of a wingless fourth-stage fuselage to which one central and seven peripheral third-stage propellant tanks have been attached. The conical disk at the nose is a radiation shield for the nuclear current generator in the tip. The concentric rings showing in the rear view (*bottom*) are manifolds for hydrazine and nitric acid and are jettisoned along with the outer tanks prior to the ship's last power maneuver.

(*Courtesy Walt Disney Productions*)

151

XL. The take-off from Mars for the circum-Martian orbit, preparatory to the return voyage to earth.

152

of the trick that makes the miracle of interplanetary radio communication possible. Of course, radioing straight across outer space is not handicapped by the curvature of the globe and the many atmospheric effects (such as dependence on ionospheric reflection, or static caused by electrical storms) which make long-range radio on earth so tricky and oftentimes unreliable. But this advantage is easily offset by the fundamental difficulty arising from the unprecedented distances which the radio waves must travel between the Mars ship and the earth. Beam or no beam, the energy impinging upon the receiver dish decreases with the square of the distance, which means that it diminishes to one-fourth as we double the distance, and to one-millionth as we increase the distance a thousandfold. Over a range of 40 million miles there is obviously very little useful signal strength left at the input of a receiver.

On the other hand, modern radio receivers perform miracles when it comes to amplifying even very weak signals. The only limitation to useful amplification lies in what radio engineers call "bandwidth." Bandwidth is a measure of the degree to which a receiver, when tuned to a certain wavelength, can also receive longer and shorter waves at the same time. The narrower the bandwidth is kept, the lower will be the background noise caused by unavoidable heat effects in the receiver tubes. Since a reasonable "signal-to-noise ratio" is required for any unblurred reception, and since the useful incoming signal strength is so weak, the fundamental requirement for interplanetary radio is to keep receiver bandwidth as narrow as possible.

Now, there are certain practical limitations set to the extent that the bandwidth can be narrowed down. First, if we are not satisfied with simple dot-and-dash communication but want to transmit music and speech, we have to provide a certain bandwidth in order to be able to receive the "overtones" with which the carrier wave is modulated and which determine the tonal quality. Next, we have to allow a certain bandwidth for the unavoidable fluctuations between the frequencies of transmitter and receiver to make sure that the transmitter does not "wander out" of the reception band for which the receiver is set. In interplanetary radio this latter problem is aggravated by the so-called "Doppler effect," which is caused by the relative motion between the ship and the orbital earth station with which it communicates. This makes the frequency actually received different from that transmitted. However, by using all the tricks modern radio technology has to offer—such as crystal-stabilized circuits and automatic frequency control—we can practically disregard frequency drift and Doppler effect, so that the bandwidth requirement is governed chiefly by the desired tonal quality of transmission.

Both the ship transmitter and the orbital station operate with only 10 kilowatts of transmitting power. While the orbital station has an antenna dish 23 feet in diameter, the ship antenna, for reasons of weight and bulk, has been reduced to 12 feet. The wavelength is 10 centimeters. The system is capable of first-rate transmission of music and speech over a distance of approximately 60 million miles, which the ships will reach about 160 days after departure. Thereafter, direct radio and voice communication will gradually cease, because the received signal strength will become so weak that the receiver bandwidth will have to be narrowed down. With automatic telegraphy, which requires much less bandwidth and which is not nearly as sensitive to background noise as voice transmission, the two-way system is capable of extending operations to a distance of more than 300 million miles. This distance is greater than that from earth to Mars at conjunction (about 235 million miles) so that communication between the ships waiting in the Martian orbit and the earth (or the space station) will never be interrupted because of distance alone. There will be an interruption when Mars, sun, and earth form a straight line with the sun in the middle. Earth will then not only be invisible optically but also out of radio contact because of the interference by the sun, which emits radio waves too. But this will be a rather short interruption of communications, lasting at most one month.

Even when the ships are not yet very far from earth, as cosmic distances go, a radio conversation will show a strange aspect. There will be pauses between question and answer because radio waves travel with the same speed as light, 186,000 miles per second. At a distance of 40 million miles the radio impulse needs 215 seconds to get from the ship to the earth station, and the reply needs 215 seconds to get back to the ship, a total of 7 minutes.

Two hundred and fifty days out.

Only ten days remain before the second major power maneuver, the induced capture of the ships by Mars. The distance to the planet—which to the naked eye looks like a reddish half-moon, one-third the diameter of earth's moon as seen from earth—has shrunk to 1,400,000 miles. The visible half-disk of Mars glows with an intense orange-red with greenish patches, and the naked eye can easily distinguish the white spot of the southern polar cap melting in the sunlight of the Martian summer. The opposite half is shrouded in night.

At a distance not much greater than the diameter of the Martian half-disk is a softly glowing starlet whose relation to Mars is visibly changing when it is observed for a few minutes. This is Phobos, Mars' inner moon, on its 7½-hour trip around the planet. Double the distance away and on the opposite side of Mars is Deimos, the other moon. Neither of the two has a diameter of more than 10 miles.

Previously the navigators were content with a "planetary parallax" measurement every 24 hours, but now they float into the astrodomes every third hour to determine the apparent motion of Mars against the fixed stars. Two days remain before the final correction maneuver is scheduled to bring the ships into the exact position from which the hyperbolic approach to the planet may begin.

The correction maneuver takes place without incident, and the velocity of both ships has been changed by 63 feet per second exactly in the direction which the navigators figured would bring the required 5470 miles' distance from the Martian orbit and simultaneously would provide proper timing with respect to the now rapidly approaching planet. A new set of observations of Mars' disk in relation to the background of fixed stars has confirmed the correctness of the outcome of the maneuver.

There is now no further use for the large propellant containers from which the power for the departure maneuver and the ensuing five corrections was drawn. The propellants still left in them are transferred to small reserve tanks, and the quick-disconnects in the propellant lines leading from the empty tanks to the pumps are released by push-button action from the control deck. Finally, the attitude-control flywheels are started to bring the two ships into slow rotation about their longitudinal axes. At the flip of a switch the explosive bolts holding the tank support struts to their sockets are detonated. The centrifugal force caused by the slow rotation causes the two pairs of great white globes to drift majestically away from the ships in opposite directions.

There are other useless loads to be dispensed with. Empty food containers, broken tools and instruments, and similar accumulated debris keep the garbage ejectors busy. Several tons of utility water, recovered by the air dryer in the air-conditioning system and hitherto used for the dishwasher, the laundry machine, and the washroom, go overboard. Every ounce of weight that can be removed contributes to saving fuel during the forthcoming power maneuver.

Events now follow in startlingly rapid succession after the weary boredom of months in interplanetary space. Twenty-four hours before the capture maneuver, Mars is four times the size of the sun, and 5 hours before the motors are to be fired, the planet appears as an enormous multicolored disk, more than half of which is illuminated by sunlight, subtending an arc of vision of more than 7 degrees— fourteen times the apparent diameter of the sun as seen from earth. The two ships, flying only a few hundred feet apart, are now only 34,000 miles from the center of Mars.

Sixty-four minutes before the rocket motors are turned on, the distance from Mars' center is clipped down to 8500 miles, and the relative speed with respect to Mars has risen from an initial 1.59 to 2.22 miles per second. Mars itself, with its glowing red, white, and green shadings, now subtends almost 30 degrees of angular vision. The planet's surface seems to rotate with ever-increasing speed, which is evidence that the ships will not crash perpendicularly upon it but are racing toward it tangentially in a graceful sweep. Now the ships are rotated into a position so that they fly tail first.

At the moment that the chronometer-controlled timer (which has been set by the guidance tape as one sets an alarm clock) fires the rocket engines, the ships' speed with respect to Mars' center is 3.20 miles per second. The acceleration is less than 1/3 g, for only two of the eight rigid-mounted rocket engines (in addition to the four hinge-mounted engines needed for flight-path control) are retarding the hyperbolic fall toward the capture orbit. The weights of the ships have been reduced so much by the consumption of propellants that the full thrust of 396 tons, with which they left the earth's orbit, would impose an excessive strain on the ships' structures and on the bodies of the crew, which have been weightless for eight months. The instant the six remaining engines (which yield a combined thrust of 132 tons) fire, the six no longer needed engines are automatically released. Since the ships are flying tail first and the operating engines are acting as brakes, the jettisoned engine package will sweep unretarded through the vertex and leave the gravitational field of Mars again on the escape leg of the capture hyperbola (see Fig. 22). The six rocket engines fire for 530 seconds, almost 9 minutes, during which time the velocity is reduced from 3.20 to 1.95 miles per second. At power cut-off the velocity is correct for a circular orbit around Mars 620 miles above the surface. This is far nearer to Mars even than Mars' very near natural moon Phobos. The period of revolution is 2 hours, 26 minutes, and 24 seconds; the plane of the orbit is the same as the plane of the orbit of the planet in its revolution around the sun.

The explorers' first task in the Martian orbit is a thorough study of the surface of the planet. The astronomical telescope enables them to discern as much detail from the 620-mile orbit as an unaided eye could see at a distance of 5000 feet. The whole planet is surveyed, photographed, and mapped; surface temperatures are measured at various latitudes by day and by night; and cloud formations are studied. All this information is immediately radioed back to the earth, to insure its preservation in case the expedition meets with disaster. Although bandwidth limitations do not permit a real television link with the earth, the radio equipment is perfectly capable of transmitting still pictures.

One of the main objectives of the minute scrutiny of the Martian surface is to select a landing place for the glider. Generally speaking, and while the expedition was still on earth, the choice of the general area was governed mostly by the understandable and logical preference for a place where a variety of Martian features are in close proximity. For temperature reasons it should not be too far from the equator, so that at least during the day artificial heat will not be needed. One promising area would be either west or east of Margaritifer Sinus, where there is a large dark area, the one just named, and several "canals"—Hydaspes and Indus—nearby. Another promising area would be to the north of Moeris Lacus, where Syrtis major is close and so is the prominent canal Thoth-Nepenthes. Because these are equatorial areas they will not only be reasonably warm during the day but will undergo little seasonal temperature changes during the stay of the expedition.

There is another reason to prefer them to the otherwise intriguing areas of Hellas in the south and the Mare acidalium in the north. A landing site near the equator of Mars quite naturally offers the same advantage as the choice of a ground base near the equator of earth, namely that the rotational velocity of the planet aids the departing ship. Mars rotates at about the same rate as the earth, but since its diameter is only about half that of earth the peripheral velocity of a point at the equator is correspondingly less. However, since the Martian gravity is much weaker too, the gain is still considerable and should be utilized, even if it may appear merely as a safety factor in the final calculations. What is more important than this gain is the fact that the plane of the ecliptic falls within a certain number of degrees of latitude to the north and south of the Martian equator, and the take-off site should be within this belt.

By the time the expedition can actually be planned, astronomical observation from the space station will have established the location of the really interesting areas. But now, with the expedition circling Mars, the choice can be narrowed further. At this short distance it will be possible to rule out a previously selected "tentative landing site" because the ground appears to have too many broken hills —*terres mauvaises*. The terrain must be flat and firm enough to permit a safe landing of the heavy craft.

Before attempting the landing, the crew conducts a thorough study of the Martian atmosphere by means of sounding missiles. These are relatively small gyroscopically controlled solid-fuel rockets which are fired in the direction opposite to the ships' orbital motion. After a short power burst which reduces the initial orbital speed by some 600 feet per second, the sounding rocket goes into an elliptical flight path whose perigee, on the opposite side of Mars, is on the fringe of the atmosphere. The drag of Martian air, increased by air brakes, slowly decelerates the missile and brings it down to the ground with ever-diminishing speed.

The sounding missiles are equipped with elaborate instrumentation which radioes a whole series of measurements back to the ship.[2] Aside from the relatively simple task of gathering and sending information on atmospheric pressure, temperature, humidity, and altitude (as determined by a radio-echo altimeter), the missiles are equipped to analyze automatically the chemical composition of the Martian atmosphere at various altitudes and to retransmit their findings by radio.

While the last telemeter messages from the sounding missiles are being evaluated, the landing craft is severed from its mother cargo ship. All equipment and supplies not needed by the landing party are removed from the large storage bin. The supplies needed for the three shipkeepers and for the final return voyage to the earth are transferred to the passenger ship. The interplanetary radio station is taken out of its well in the walls of the storage bin and temporarily connected to the hull of the passenger ship. The astronomical telescope is simply left floating in the orbit. It has its own ultra-precise spatial attitude-control system and will be used by the shipkeepers to observe the progress of the landing party.

The three men who have the somewhat thankless and certainly monotonous assignment of staying in the orbit around Mars while the expedition proper descends to the surface are under strictest orders to return to earth when the waiting time has elapsed, even if the ground expedition has not returned. Their job is, very simply and somewhat cynically, to get some results from the expedition even if something should go wrong on Mars.

2. This "telemetering" technique is widely used "in reverse" in today's upper-atmosphere rockets.

It has been said that every major mishap that befalls an expedition is, in the last analysis, due to poor or incomplete planning. This may be perfectly true for expeditions to the polar regions of earth, or for expeditions to Mount Everest, because in these cases there is enough previous experience to make complete and perfect planning theoretically possible. But in regard to the first expedition to another planet no such statement can be made. Everything foreseeable will have been foreseen, every emergency conceivable to terrestrial experience will have been considered, but since this is not earth some things may not be foreseeable. In short, it is possible, though rather unlikely, that something will happen to the ground expediton. But since the ground expedition will report to the orbiting ship whenever there is something to report—and will make routine reports at regular intervals even when there is nothing special to report—at least all the information gathered will be recorded on the orbiting ship. And if something should go wrong on the ground, the reason would probably be known from these reports. The men of the orbiting ship act as a temporary receptacle for all information gathered, and it is their duty to see that it gets back to earth.

After a few more days of painstaking inspection of the landing craft and its surface cargo, the nine explorers strap themselves into their seats for the descent to Mars. Ponderously the great glider cartwheels under the effect of the side thrusts from its small attitude-control nozzles, until it is coasting through the orbit tail first. Then the small landing motor fires. In 157 seconds of burning time, its weak thrust of a mere 22 tons decelerates the 177-ton glider by as little as 565 feet per second. When the engine finally cuts out, free-coasting flight through the landing ellipse begins. In a little over an hour, this landing ellipse takes the glider halfway around Mars' multicolored globe and into the Martian night. A hissing sound, hardly audible at first but growing in strength, indicates that the glider has entered the Martian atmosphere, and soon the craft begins to respond to the pilot's handling of the airplane-type controls. Soon thereafter the electric altimeter indicates that the glider has reached the perigee of the landing ellipse, at an altitude of 96 miles.

The enormous altitude at which the decelerated aerodynamical glide commences is rather typical of the queer physical make-up of the red planet. It seems to be in direct contradiction to the low density of the Martian atmosphere, and yet it is a perfectly logical choice. For the low air density on Mars' surface (corresponding to that in the earth's atmosphere at an altitude of 11 miles) is only in part due to the fact that there is actually less air above the Martian surface. An equally important factor is the feeble gravitational field of Mars which is unable to compress the atmosphere into such a thin layer as the atmosphere of earth. As a result, Mars' atmosphere, tenuous as it is, is actually higher than the earth's atmospheric shell, and above altitudes of about 18 miles even the absolute atmospheric density exceeds that of the earth's atmosphere.[3]

The speed of 2.28 miles per second at perigee slowly diminishes under the effect of air drag. At first, the explorers will feel a slight sensation of hanging in their shoulder belts, as the pilot puts his stick forward to prevent the craft from coasting out of the atmosphere again on the second leg of the landing ellipse. But as the velocity approaches circular speed the sensation subsides, and they feel only a slight forward pull, indicating continued retardation by air drag. Finally the previous effect reverses itself, and they begin, for the first time in nine months, to feel their weight, slightly at first, but to an increasing extent as the lift of the wings is called upon to prevent the glider from descending too rapidly into the deeper layers of the Martian atmosphere. The night sky brightens, turning from purple to blue (the first blue sky for months!), and finally the glider emerges into sunlight again. At an altitude of 24 miles, the glider, having gone more than halfway around Mars since passing the perigee, has a velocity only of the speed of sound. At subsonic speed it now spirals down to the predetermined landing spot, with the nose flaps lowered to improve the aerodynamic qualities of the thin supersonic wings at low air speeds. As has been mentioned at the end of Chapter 6, it would be unwise to consider a high landing speed permissible, especially since the landing craft must land on unprepared terrain. A landing speed of 120 miles per hour is probably the highest one can safely allow. This, considering the thinness of the Martian atmosphere, accounts for the very large wings. The lack of concrete runways on Mars is also reflected in the heavy "bicycle"-type landing gear, consisting of two wide skids into which caterpillar tracks have been imbedded. The forward skid extends from the craft's nose; the rear one is located under the thickened aft section of the hull. Two additional light outrigger skids have been provided to support the wings after the craft has come to a stop.

When the landing craft has reached an altitude of only a few thousand feet above the ground, a smoke bomb is dropped to determine the wind direction and

3. This is also the astronomers' explanation for the repeated observations of clouds in the Martian atmosphere at altitudes much higher than any normal clouds on earth reach.

the best course for the landing approach. Flaps, track landing gear, and outrigger skids are lowered, and finally the heavy craft, touching down at 120 miles per hour, rumbles over the sandy plains and grinds to a stop in a billowing cloud of dust (Plate xxx).

Clad in pressurized suits, the first nine human beings to set foot on Mars are grouped around the cabin door. One by one they enter the airlock and listen to the hiss of the escaping air as the lock is brought down to the low pressure outside. The outer door opens and they step out onto the huge wing.

The scene before them might well be a desert region in the American southwest, glistening in the sunlight under a dark blue, cloudless sky.

The sensation of restored gravity is most unpleasant. The men feel weak-kneed after more than nine months of weightlessness. Only as they reluctantly take the jump of 18 feet separating the wing's leading edge from the ground do they realize how feeble the gravitation actually is, for the fall is not more than a gentle floating down to the sand. Nevertheless, they feel as though they had lead in their veins and have an urge to lie down. But there's no time for a rest now. As intruders on a strange planet which might have all kinds of surprises in store, they cannot afford to take any chances.

The first task is to ready the nose section of the landing craft in order to be ready for a hasty retreat to the circum-Martian orbit. The large bottom hatch of the cargo compartment is opened, and the two caterpillar tractors are lowered to the ground. Two struts are attached to the upper side of the torpedo-like nose so that they form a kind of bridge. A steel cable is strung from the apex of this bridge to the winch of one caterpillar stationed behind the stern of the landing craft (see Plate xxxi). Another cable is run from the tip of the rocket's nose to the second caterpillar positioned forward of the craft. The two large wings are jacked up near their roots to provide a firm support for the erection of the nose portion. Finally the bolts connecting the nose and the aft portion of the landing craft are removed and the difficult process of erection begins.

As the rearward tractor starts winding in the one cable and the forward caterpillar slackens the other, the nose section hinges around the two pivots hidden inside the wing roots. A few minutes later the rocket is in vertical position, its fins resting on broad aluminum shoes placed on the sand.

The next job is the preparation of the pneumatic tent that will be home and headquarters for the stay on Martian soil. The "tent"—actually a hemispherical dome about 20 feet in diameter—is made of rubberized fabric. It is padded with effective heat insulation, and had been stowed in the landing craft's cargo room collapsed, like an oversize rubber raft. After one of the caterpillars has hauled it to a suitable position not far from the landing craft, it is inflated with the standard "space man's atmosphere"—40 per cent oxygen and 60 per cent helium, at a pressure of 8 pounds per square inch. In this tent the explorers can eat, sleep, and work without the pressurized discomfort of the tractors and space suits, in air-conditioned luxury. The airlock through which it may be entered and left provides storage space for the pressure suits and is equipped with germ-killing radiation lamps to ward off any hazards from potentially dangerous Martian bacteria.

The tractors provide ground transportation for the exploration of the terrain within about a hundred miles around our landing site. Like all other equipment, they have been designed to be as light as possible, but with all their complicated installations they still weigh 3.3 earth tons apiece. On Mars, however, their actual weight is only a little over a ton. Therefore, both vehicles had to be anchored securely to the ground while the winches were erecting the heavy rocket.

Each tractor's cylindrical, pressurized body provides space for four men and ample storage space for supplies and research equipment. At the rear there is a small crane of the type used in tow cars, its functions being combined with those of the winch.

The motor power of the tractors can take several forms, depending on the availability of small and reliable atomic reactors at the time. If suitable atomic reactors exist by then they might conceivably produce steam for a closed-cycle turbine drive, in about the same fashion as does the power plant of the submarine *Nautilus*. The advantage of a nuclear reactor would be that the fuel weight is virtually zero, even if the installation itself might weigh more than the fuel-driven engine to be described. An atomically driven tractor would have an almost unlimited range, which unfortunately is not true for the tractor that could be designed now.

In this design the engine is an especially interesting feature; because of the lack of oxygen in the Martian atmosphere it uses two propellants as a rocket motor does—namely, hydrogen peroxide and fuel oil. In a "catalyst chamber" the hydrogen peroxide is first split chemically into a mixture of high-temperature steam and free oxygen. Next, fuel oil is injected, which burns in the oxygen of the mixture.

162

Finally, water is sprayed into the flame. The result is steam of moderate temperature, only slightly contaminated by carbon dioxide and carbon monoxide from the combustion of the oil. The flow of this steam can be regulated by throttling of its three constituents. It turns a turbine which provides power for the caterpillar tracks. After passing through the turbine the steam is condensed in a pair of low-pressure radiators cooled by the thin Martian air with the aid of two large blowers. The carbon dioxide and carbon monoxide remain in the vapor phase and are drawn off and expelled by a small compressor, while the water in liquid phase is recirculated from the radiators to the combustion chamber.

The 6.6 tons of tractor propellants brought to Mars as part of the landing craft's cargo are sufficient for a total mileage for one tractor of around 1600 miles. Since both tractors are likely to be used to about the same extent one may say that each tractor can be driven 800 miles. The capacity of the tractor's propellant tanks is such that a 50- or 60-mile run can be made without refueling, but this range can be extended, if necessary, by hooking a fuel trailer to the tractor.

After more than a year on Martian soil, the supplies are nearly exhausted, and the explorers begin to prepare for departure. The year has been one of privation and tensions comparable to those of a winter camp in the Arctic. The party has had to deal with mechanical difficulties and all kinds of surprises which the designers of the equipment could not possibly have expected. Dust storms (Plate XXXVIII) threatened the intricate machinery of the return rocket, causing the explorers to fear that they might be permanently grounded on this strange planet.

But it has also been a year of exciting scientific discoveries, of profound satisfaction in being able to study meteorology and climate, rock formations and soil bacteria, plant life and seasonal changes on another planet. The explorers have sounded out the internal make-up of Mars by detonating explosive charges on the ground and measuring the propagation of the shock waves with seismographs planted several miles around the blast. They have searched for possible remnants of higher forms of life that might have populated Mars in past geological ages, and for indications of whether Mars has ever been inhabited by intelligent beings.

All findings of every kind have been carefully reported to the passenger ship in the orbit, and the shipkeepers have relayed the information to the earth. Thousands of color photographs as well as a substantial footage of motion-picture film will be carried back, as well as a vast collection of minerals and specimens of Martian plant life, limited only by the payload restrictions of the rocket ship in which the party will return.

The take-off is precisely timed in accordance with the latest orbital data from the circling passenger ship. During the nearly 400 days of the stay on Mars the orbit has been subjected to certain perturbations which must be taken into account.

The last week on Mars is spent in a thorough check of all components of the rocket. At last the hour of departure comes. With an acceleration of about 1 g the rocket rises from the Martian sands (Plate xxxix). Soon the flight track is tilted into an easterly course. After 147 seconds of burning time the rocket reaches a speed of 2.3 miles per second, flying exactly horizontally. A little over an hour later it reaches the apogee of the unpowered ellipse of ascent. In a very short burst of power, with the rocket engine at half-throttle only, its speed is matched with that of the orbiting passenger ship. Another critical hurdle has been passed.

Through the circular port of the rocket the explorers see the huge deep-space ship hanging without apparent motion against the black, star-studded heavens. Its appearance has changed during their absence. The large cylindrical tanks containing the propellants for the capture maneuver are gone, and only the cylindrical central body, housing the tanks for the return fuel, is left. The interplanetary radio station has been detached from the passenger ship and floats nearby. The odd structure of tanks and rocket engines forming the "interplanetary booster" for the landing craft is gone. Because of slight differences in the orbital data it has slowly drifted away from the passenger ship. It will remain in its orbit forever as a third Martian moon.

After donning space suits the men transfer to the passenger ship and experience once more the thrill of floating freely a few hundred miles above Mars' magnificent, multicolored landscape.

"X minus one minute" resounds from the bullhorns.

The final test has come. Will the complicated mechanism of the six-engine rocket power plant, the vast array of gyroscopes, computers, switches, actuators, and relays that make up the guidance and control system, still be in the same perfect condition that they were when the ship swept into the orbit of Mars 449 days ago? Hundreds of checks and functional tests have allegedly removed all doubts in the minds of the men who remained with the ship, but perhaps the shipkeepers have been overtaxed by the strain. Within a few seconds the answer will be known.

"X minus twenty seconds."

Co-pilot and engineer rapidly scan their complicated instrument panels. The whine of the gyroscopes penetrates the monotonous rustle of the respiration blower. Now a click-click-click indicates that the flight program tape is running.

"X minus ten—nine—eight—seven—six—five—four—"

Ignition stage and tense waiting. A few seconds later the main engines, at main stage, roar their deep-throated song of power. After a little over 4 minutes, cut-off and silence. The return trip has begun.

At last the grueling 260-day trip back to earth is nearing its end; preparation for the final capture maneuver into the return orbit begins. A final correction maneuver puts the ship into a position 79,000 miles outside of the earth's orbit. The earth itself is still a million miles ahead, but the ship is moving faster than the earth and begins the hyperbolic fall into the earth's gravitational field with an excess speed of 1.88 miles per second. Slowly the velocity begins to rise. The ship cartwheels around to fly tail first. With a speed of 2.65 miles per second relative to the earth, it sweeps toward the vertex of the capture hyperbola. Only the four hinge-mounted rocket engines fire. At the moment of ignition, the last two rigid engines are detached and sweep unretarded through the vertex of the hyperbola and out into interplanetary space again.

After 306 seconds of burning time the roar subsides. The speed is down to 1.32 miles per second. The ship has settled in the return orbit, 56,000 miles from the earth's center.

A few days after the capture maneuver the special relief ship of the Space Lift organization sweeps into the orbit and takes the members of the expedition down to the altitude of the departure orbit, only 1075 miles above the earth's surface. Here they are once more transferred—this time to a winged personnel stage of one of the Space Lift's orbital supply ships. After an hour of coasting through the landing ellipse the hissing of the atmosphere becomes audible, and another hour later the ship touches down on the airstrip of the base from which the explorers took off two and one-half years before.

The first Mars expedition has ended.

TABLES

TABLE I. MAIN PERFORMANCE DATA OF ORBITAL SUPPLY SHIP

	U.S.	METRIC
Orbit of departure		
Period of revolution	2 hr 0 min 0 sec	
Orbital radius	5040 mi	8110 km
Orbital altitude	1075 mi	1730 km
Orbital velocity	4.40 mi/sec	7.07 km/sec
Inclination of orbital plane against equatorial plane	23.5°	
First stage		
1 Thrust, at expansion to nozzle exit pressure	2810 tons	2560 t
2 Take-off weight	1410 tons	1280 t
3 Empty weight	154 tons	140 t
4 Final weight (1st stage empty, 2nd and 3rd full)	352 tons	320 t
5 Ratio of empty weight of 1st stage to propellant weight	0.146	
6 Propellant weight	1058 tons	960 t
7 Rate of propellant flow	12.3 tons/sec	11.15 t/sec
8 Exhaust velocity	7400 ft/sec	2250 m/sec
9 Specific impulse (sea level)	230 sec	
10 Take-off acceleration (relative)	0.9 g	
11 Final acceleration (absolute)	8.9 g	
12 Burning time	84 sec	
13 Burning time reserve	2 sec	
14 Cut-off altitude	24.9 mi	40 km
15 Cut-off velocity	1.46 mi/sec	2350 m/sec
16 Cut-off distance	31.1 mi	50 km
17 Angle of elevation at cut-off	20.5°	
18 Length of 1st stage (without fins, but incl. well for 2nd-stage engines)	73.0 ft	22.2 m
19 Base diameter of 1st stage	38.4 ft	11.7 m
Second stage		
20 Thrust	352 tons	320 t
21 Initial weight	198 tons	180 t
22 Empty weight	15.4 tons	14 t
23 Final weight (2nd stage empty, 3rd stage full)	44.0 tons	40 t
24 Ratio of empty weight of 2nd stage to propellant weight	0.10	
25 Propellant weight	154 tons	140 t
26 Rate of propellant flow	1.23 tons/sec	1.12 t/sec
27 Exhaust velocity	9200 ft/sec	2800 m/sec
28 Specific impulse	285 sec	
29 Initial acceleration (absolute)	1.8 g	
30 Final acceleration (absolute)	8.0 g	
31 Burning time	124 sec	
32 Burning time reserve	1 sec	
33 Cut-off altitude	39.8 mi	64 km
34 Cut-off velocity	3.99 mi/sec	6420 m/sec
35 Cut-off distance (from take-off)	332 mi	534 km
36 Angle of elevation at cut-off	2.5°	
37 Length of 2nd stage (without engines, but incl. well for 3rd-stage engine)	55.5 ft	16.9 m
38 Base diameter of 2nd stage (tanks)	25.6 ft	7.8 m
Third stage (maneuver of ascent)		
39 Thrust	44.0 tons	40 t
40 Initial weight	28.6 tons	26 t
41 Empty weight without payload nose	2.3 tons	2.1 t
42 Final weight after maneuver of ascent	17.28 tons	15.7 t
43 Ratio of empty weight (#41) to propellant weight (# 44 + 60)	0.165	
44 Propellant weight (maneuver of ascent)	11.32 tons	10.3 t
45 Rate of propellant flow	310 lb/sec	141 kg/sec
46 Exhaust velocity	9200 ft/sec	2800 m/sec
47 Specific impulse	285 sec	
48 Initial acceleration (absolute)	1.54 g	
49 Final acceleration (absolute)	2.55 g	
50 Burning time	73 sec	
51 Cut-off altitude	63.3 mi	102 km
52 Cut-off velocity, incl. contribution of 1393 ft/sec (425 m/sec) from earth rotation	5.13 mi/sec	8260 m/sec
53 Cut-off distance (from take-off)	655 mi	1054 km
54 Angle of elevation at cut-off	0°	
55 Length of 3rd stage (without engine, # 37, payload nose or tank extension for propellant supply flights, see # 72)	9.5 ft	2.9 m
56 Base diameter of 3rd stage (tanks)	7.0 ft	2.14 m
Third stage (maneuver of adaptation in orbit)		
57 Thrust	44.0 tons	40 t
58 Initial weight	17.28 tons	15.7 t
59 Final weight after adaptation maneuver	14.64 tons	13.3 t
60 Propellant weight	2.64 tons	2.4 t
61 Burning time	17 sec	
62 Velocity increment for adaptation	0.286 mi/sec	460 m/sec
Weight breakdown of third stage		
63 Rocket engine—44 tons thrust, swivel-mounted	1.10 tons	1.0 t
64 Propellant tanks, incl. pressurization;* capacity 14 tons	0.77 tons	0.7 t
65 Swivel actuators	0.12 tons	0.1 t
66 Roll and spatial attitude-control nozzles	0.33 tons	0.3 t
67 Total dry weight	2.32 tons	2.1 t
Weight breakdown of payload nose (unmanned)		
68 Total weight (# 59 —67)	12.32 tons	11.2 t
69 Guidance and control equipment, incl. battery power supply, distributors, radio remote-control receiver for adaptation maneuver, etc.	1.10 tons	1.0 t
70 Cargo bin for dry cargo	0.22 tons	0.2 t
71 Available dry cargo weight (# 68 —69 —70)	11.0 tons	10.0 t
72 Available propellant cargo weight*	11.2 tons	10.2 t

* Propellants for the Mars ships are shipped to the orbit in extra-long 3rd-stage propellant tanks of 25.2 tons (22.9 t) capacity. This is 11.2 tons (10.2 t) more than the tank capacity of 14.0 tons (12.7 t) used for "dry cargo" and "passenger" flights. In spite of their greater capacity these tanks are not heavier, since they need not carry the longitudinal and bending loads imposed by the "dry payload nose" or the earth-returnable personnel stage.

	U.S.	METRIC
Weight breakdown of earth-returnable nose section (manned 4th stage)		
73 Total weight at departure from orbit (same as # 68)	12.32 tons	11.2 t
74 Propellants for return maneuver (see # 96)	1.98 tons	1.8 t
75 Final weight (landing weight) (# 73 —74)	10.34 tons	9.4 t
76 Payload, to or from orbit: 14 men, incl. crew of 2, plus 1.1 tons extra cargo	2.64 tons	2.4 t
77 Hull (heat-protected for hypersonic glide)	0.88 tons	0.8 t
78 Wing (ditto)	1.54 tons	1.4 t
79 Empennage (ditto)	0.33 tons	0.3 t
80 Landing gear	0.55 tons	0.5 t
81 Seats	0.11 tons	0.1 t
82 Airlock	0.33 tons	0.3 t
83 Portholes and astrodome	0.33 tons	0.3 t
84 Guidance and control equipment for ascent (see # 69)	1.10 tons	1.0 t
85 Remote-control equipment for adaptation guidance of unmanned supply ships	0.22 tons	0.2 t
86 Instrumentation for descent: flight instruments, radio, etc.	0.44 tons	0.4 t
87 Spatial attitude-control nozzles	0.33 tons	0.3 t
88 Heat insulation for cabin (hypersonic glide)	0.22 tons	0.2 t
89 Air-conditioning system, incl. oxygen supply for 5 days	1.10 tons	1.0 t
90 Rocket engine for return maneuver—1.1 tons thrust	0.11 tons	0.1 t
91 Propellant tanks for return maneuver	0.11 tons	0.1 t
92 Total (# 76 through 91, equal to 75)	10.34 tons	9.4 t

	U.S.	METRIC
Earth-returnable, manned 4th stage (maneuver of return, glide, and landing)		
93 Thrust	1.1 tons	1.0 t
94 Initial weight (same as # 68, 73)	12.32 tons	11.2 t
95 Final weight (landing weight)	10.34 tons	9.4 t
96 Propellant weight	1.98 tons	1.8 t
97 Rate of propellant flow	7.70 lb/sec	3.5 kg/sec
98 Burning time	515 sec	
99 Velocity decrement for return maneuver	0.298 mi/sec	480 m/sec
100 Altitude of perigee of landing ellipse	49.7 mi	80 km
101 Wing area	1385 sq ft	129 m²
102 Wing loading	15 lb/sq ft	73 kg/m²
103 Maximum wing temperature during hypersonic glide †	1350° F	732° C
104 Altitude at which air speed is down to sonic speed	15 mi	24 km
105 Landing speed	65 mi/hr	105 km/hr
Dimensions of orbital supply ship		
106 Base diameter (aft end of 1st stage)	38.4 ft	11.7 m
107 Length from aft base 1st stage to front end 3rd-stage tanks (# 18 + 37 + 55)	138.0 ft	42.0 m
108 Length of payload nose (dependent upon cargo)	26–53 ft	8–16 m
109 Length of earth-returnable, manned 4th stage	42.8 ft	13 m
110 Wing span of same	83 ft	25.4 m
111 Diameter of payload noses (manned and unmanned)	7.0 ft	2.14 m
112 Over-all length of supply ships		
Unmanned (# 107 + 108)	164–191 ft	50–58 m
Manned (# 107 + 109)	180.8 ft	55 m

† Calculations, based on assumption of turbulent flow, refer to a point 1 ft downstream from leading edge. Maximum temperature occurs at a speed of approximately 3.1 mi/sec (5 km/sec) and an altitude of 37 miles (60 km).

TABLE II. MAIN PERFORMANCE DATA OF MARS SHIPS

(Cargo ship performs maneuvers 1 and 2 only)

	U.S.	METRIC
Maneuver 1—departure from 2-hr orbit		
1 Thrust: 8 rigid engines, 44 tons (40 t) thrust each; 4 hinged engines, 11 tons (10 t) thrust each	396.0 tons	360.0 t
2 Rate of propellant flow	2772 lb/sec	1260 kg/sec
3 Initial weight	1870.0 tons	1700.0 t
4 Propellant supply (incl. 10 % velocity reserve)	1370.3 tons	1246.6 t
5 Final weight (tanks maneuver 1 exhausted)	499.7 tons	453.4 t
6 Initial acceleration	0.212 g	
7 Final acceleration (referred to actual, not total, burning time)	0.710 g	
8 Actual burning time	948 sec	
9 Initial orbital velocity (with respect to earth's center)	4.40 mi/sec	7.07 km/sec
10 Radius of departure orbit	5040 mi	8110 km
11 Cut-off velocity (with respect to earth's center)	5.99 mi/sec	9.64 km/sec
12 Geocentric angle covered during maneuver 1	55.4°	
13 Velocity increment maneuver 1 (# 11 —9)	1.59 mi/sec	2.57 km/sec
14 Distance of cut-off point from earth's center	5930 mi	9550 km

	U.S.	METRIC
15 Residual velocity after escape from earth's gravitational field (with respect to earth's center)	1.88 mi/sec	3.03 km/sec
16 Orbital velocity of earth (with respect to sun)	18.52 mi/sec	29.80 km/sec
17 Perihelion velocity of Mars-bound voyaging ellipse (# 15 + 16)	20.40 mi/sec	32.83 km/sec
18 Total burning time to tank exhaustion	990 sec	
19 Velocity reserve for flight path corrections after maneuver 1	0.19 mi/sec	0.30 km/sec
Weight reduction prior to maneuver 2		
20 Oxygen, food, and water consumed by 6-man crew during 260-day flight *	—8.1 tons	—7.4 t

(Continued on next page)

* Because of the limited crew spaces in the landing craft, only 4 crew members ride in the cargo ship, whereas 8 travel in the passenger ship. In order to keep initial weights of both ships alike for maneuver 2, supplies for 2 men (2.64 tons, 2.4 t) are transferred from cargo to passenger ship during 260-day Mars-bound flight.

TABLE II (continued)

	U.S.	METRIC
21 4 tanks for maneuver 1 detached (incl. supports, pressurization system, propellant lines, and 2-mm aluminum meteor bumpers for tanks)	−5.0 tons	−4.5 t
22 6 rigid rocket engines detached	−6.6 tons	−6.0 t
23 Total weight reduction	−19.7 tons	−17.9 t

Maneuver 2—capture in Mars orbit

	U.S.	METRIC
24 Thrust: 2 rigid engines, 44 tons (40 t) thrust each; 4 hinged engines, 11 tons (10 t) thrust each	132.0 tons	120.0 t
25 Rate of propellant flow	926 lb/sec	421 kg/sec
26 Initial weight (# 5 − 23)	480.0 tons	435.5 t
27 Propellant supply, incl. 10 % velocity reserve	262.0 tons	237.5 t
28 Final weight (maneuver 2 tanks exhausted)	218.0 tons	198.0 t
29 Initial acceleration	0.275 g	
30 Final acceleration (referred to actual, not total, burning time)	0.564 g	
31 Actual burning time	530 sec	
32 Aphelion velocity of Mars-bound ellipse (with respect to sun)	13.39 mi/sec	21.55 km/sec
33 Orbital velocity of Mars	14.98 mi/sec	24.10 km/sec
34 Initial velocity of hyperbolic fall toward Mars (with respect to Mars' center; # 33−32)	1.59 mi/sec	2.55 km/sec
35 Distance of asymptote of capture hyperbola from Mars' center	5470 mi	8800 km
36 Velocity at vertex of capture hyperbola (with respect to Mars' center)	3.20 mi/sec	5.15 km/sec
37 Orbital velocity of capture orbit (with respect to Mars's center), altitude 620 mi (1000 km)	1.95 mi/sec	3.14 km/sec
38 Velocity decrement maneuver 2 (# 36 −37)	1.25 mi/sec	2.01 km/sec
39 Total burning time to tank exhaustion	564 sec	
40 Velocity reserve for orbit corrections after maneuver 2	0.12 mi/sec	0.20 km/sec

Weight increase prior to maneuver 3

	U.S.	METRIC
41 Oxygen, food, and water for 260-day return flight plus 20-day reserve (contact by relief ship) for 12-man crew. (These supplies are transferred from cargo to passenger ship during "waiting time.")	+17.22 tons	+15.7 t
42 2 tanks for maneuver 2 detached, incl. supports, pressurization system, propellant lines, and 2-mm aluminum meteor bumpers for tanks	−4.06 tons	−3.7 t
43 4 more crew members and 5.5 tons research specimens from Mars taken aboard	+5.94 tons	+5.4 t
44 Total weight increase	+19.10 tons	+17.4 t

Maneuver 3—departure from Mars orbit

	U.S.	METRIC
45 Thrust: 2 rigid engines, 44 tons thrust each; 4 hinged engines, 11 tons thrust each	132.0 tons	120.0 t
46 Rate of propellant flow	926 lb/sec	421 kg/sec
47 Initial weight (# 28 + 44)	237.1 tons	215.4 t
48 Propellant supply (incl. 10 % velocity reserve)	129.3 tons	117.3 t
49 Final weight (maneuver 3 tanks exhausted)	107.8 tons	98.1 t
50 Initial acceleration	0.558 g	
51 Final acceleration (referred to actual, not total, burning time)	1.144 g	
52 Actual burning time	262 sec	
53 Initial orbital velocity (with respect to Mars' center)	1.95 mi/sec	3.14 km/sec
54 Cut-off velocity (with respect to Mars' center)	3.20 mi/sec	5.15 km/sec
55 Velocity increment maneuver 3 (# 54 − 53)	1.25 mi/sec	2.01 km/sec
56 Distance of asymptote of escape hyperbola from Mars' center	5470 mi	8800 km
57 Residual velocity after escape from Mars' gravitational field (with respect to Mars' center)	1.59 mi/sec	2.55 km/sec
58 Orbital velocity of Mars	14.98 mi/sec	24.10 km/sec
59 Aphelion velocity of earth-bound voyaging ellipse (# 58 −57)	13.39 mi/sec	21.55 km/sec
60 Total burning time to tank exhaustion	279 sec	
61 Velocity reserve for flight path corrections after maneuver 3	0.12 mi/sec	0.20 km/sec

Weight reduction prior to maneuver 4

	U.S.	METRIC
62 Oxygen, food, and water consumed by 12-man crew during 260-day flight (20-day reserve for wait for relief ship still left)	−16.1 tons	−14.6 t
63 2 rigid rocket engines detached	−2.2 tons	−2.0 t
64 Total weight reduction	−18.3 tons	−16.6 t

Maneuver 4—return into 74-hr earth orbit

	U.S.	METRIC
65 Thrust: 4 hinged engines, 11 tons (10 t) thrust each	44.0 tons	40.0 t
66 Rate of propellant flow	308 lb/sec	140 kg/sec
67 Initial weight (# 49 −64)	89.5 tons	81.5 t
68 Propellant supply (incl. 10 % velocity reserve)	51.1 tons	46.6 t
69 Final weight (maneuver 4 tanks exhausted)	38.4 tons	34.9 t
70 Initial acceleration	0.491 g	
71 Final acceleration (referred to actual, not total, burning time)	1.036 g	
72 Actual burning time	306 sec	
73 Perihelion velocity of earth-bound voyaging ellipse (with respect to sun)	20.40 mi/sec	32.83 km/sec
74 Orbital velocity of earth	18.52 mi/sec	29.80 km/sec
75 Initial velocity of hyperbolic fall toward earth (with respect to earth's center, # 73 −74)	1.88 mi/sec	3.03 km/sec

	U.S.	METRIC		U.S.	METRIC
76 Distance of asymptote of capture hyperbola from earth's center	79,000 mi	127,000 km	102 12 space suits	1.32 tons	1.2 t
			103 Spares and tools	2.10 tons	1.9 t
			104 12 crew members	1.32 tons	1.2 t
77 Velocity at vertex of capture hyperbola (with respect to earth's center)	2.65 mi/sec	4.26 km/sec	105 Personal baggage	1.32 tons	1.2 t
78 Orbital velocity of capture orbit (with respect to earth's center), orbital radius 56,000 mi (90,000 km)	1.32 mi/sec	2.12 km/sec	106 20-day oxygen, food, and water reserve for contact with relief ship	1.20 tons	1.1 t
79 Velocity decrement maneuver 4 (# 77 —78)	1.33 mi/sec	2.14 km/sec	107 Research specimens collected on Mars	5.50 tons	5.0 t
80 Total burning time to tank exhaustion	333 sec		108 Payload total	12.76 tons	11.6 t
81 Velocity reserve for corrections after maneuver 4	0.13 mi/sec	0.21 km/sec	109 Grand total (see # 69)	38.40 tons	34.9 t

Weight breakdown of passenger ship after return into the earth orbit

Construction weight (dry weight) of passenger ship prior to departure from earth orbit

	U.S.	METRIC		U.S.	METRIC
82 Shell of spherical crew nacelle: sphere of 26 ft (8 meters) diameter; internal pressure 8 psi (0.56 kg/cm²); rubber-impregnated nylon, covered with 2-mm aluminum meteor bumper	2.75 tons	2.5 t	110 Dry ship weight after return into earth orbit, maneuver 4 (see # 101)	25.64 tons	23.3 t
83 Airlock	0.55 tons	0.5 t	111 2 engines detached prior to maneuver 4 (# 63)	2.20 tons	2.0 t
84 Portholes and astrodome	0.66 tons	0.6 t	112 2 tanks detached prior to maneuver 3 (# 42)	4.06 tons	3.7 t
85 Furnishings (flooring, food-storage compartment, "dining unit," garbage ejector, toilet, washing machine, etc.)	1.87 tons	1.7 t	113 4 tanks detached prior to maneuver 2 (# 21)	5.00 tons	4.5 t
86 Guidance equipment	2.20 tons	2.0 t	114 6 engines detached prior to maneuver 2 (# 22)	6.60 tons	6.0 t
87 Navigational equipment	0.77 tons	0.7 t	115 Dry ship weight prior to maneuver 1	43.50 tons	39.5 t
88 Combined water and air recovery system, incl. liquid oxygen and helium storage tanks	2.75 tons	2.5 t			

Weight breakdown of cargo ship after arrival in the circum-Martian orbit

	U.S.	METRIC		U.S.	METRIC
89 Radio for moderate range, incl. antennae	0.88 tons	0.8 t	116 Arrival weight in circum-Martian orbit (see # 28)	218.0 tons	198.0 t
90 Nuclear electrical power source (20 kw), incl. radiation shield	2.10 tons	1.9 t	117 2 tanks for maneuver 2 (see # 42)	4.06 tons	3.7 t
91 Emergency electrical power source (5 kw)—solar and chemical batteries	0.88 tons	0.8 t	118 2 rigid rocket engines (see # 63)	2.20 tons	2.0 t
92 Instrumentation, temperature controls	0.88 tons	0.8 t	119 4 hinge-mounted rocket engines (see # 99)	1.54 tons	1.4 t
93 Cylindrical ship's structure: cylinder of 26 ft (8 m) diameter and 59 ft (18 m) length. Weight based on a shell structure of 2-mm aluminum	2.75 tons	2.5 t	120 Hydraulic actuators (see # 100)	0.44 tons	0.4 t
94 Tanks for maneuvers 3 and 4 (plastic containers, no meteor bumpers required since tanks are located inside cylindrical shell)	0.88 tons	0.8 t	121 Weight of "interplanetary booster" (# 117 through 120)	8.24 tons	7.5 t
95 Pressurization system for these tanks (helium flasks and reducers to provide a tank pressure of 5 psi)	0.55 tons	0.5 t	122 Fully loaded and fueled landing craft, incl. cargoes not needed on Mars' surface (# 116 —121)	209.76 tons	190.5 t
96 Propellant lines to engines, vent valves, etc.	0.66 tons	0.6 t	123 Initial weight of landing craft prior to descent	177.0 tons	161.0 t
97 Spatial attitude-control system (flywheels and controls)	1.43 tons	1.3 t	124 Cargo removed from landing craft's cargo bin prior to descent to Mars' surface (# 122 —123; consists of # 125 through 130)	32.76 tons	29.5 t
98 Cable harness, distributors, etc.	1.10 tons	1.0 t	125 Oxygen, food, and water for 260-day return flight plus 20-day reserve (contact by relief ship) for 12-man crew (see # 41)†	17.22 tons	15.7 t
99 4 hinge-mounted rocket engines, thrust 11 tons each. (Weight includes hinge mounts, thrust chambers, turbopumps, gas generator, valves, and internal piping)	1.54 tons	1.4 t	126 Oxygen, food, and water for 449-day "waiting time" for 3 shipkeepers who remain in passenger ship †	7.00 tons	6.3 t
100 Hydraulic actuators (for deflection of rocket engines)	0.44 tons	0.4 t	127 Oxygen, food, and water for 49 days, during which 9 members of landing party stay in circum-Martian orbit (see also Table III # 12)†	2.40 tons	2.1 t
101 Ship total	25.64 tons	23.3 t	128 Telescope	1.80 tons	1.5 t
			129 Interplanetary radio station	2.20 tons	2.0 t
			130 Sounding missiles, incl. launching gear and telemeter receiving equipment	2.14 tons	1.9 t

† # 125, 126, and 127 are transferred to passenger ship prior to descent to Mars' surface.

TABLE III. MAIN PERFORMANCE DATA OF LANDING CRAFT

		U.S.	METRIC
Maneuver of descent in circum-Martian orbit			
1	Thrust	22.0 tons	20.0 t
2	Initial weight	177.0 tons	161.0 t
3	Final weight ("earth weight")	165.0 tons	150.0 t
4	Burning time, incl. 1.1 tons propellant reserve	157 sec	
5	Velocity decrement for maneuver of descent	565 ft/sec	173 m/sec
6	Altitude of perigee of landing ellipse	96.5 mi	155 km
7	Landing speed	122 mi/hr	196 km/hr
Weight reduction prior to re-ascent			
8	Wings (wing area 24,500 sq ft, 2280 m²)	27.00 tons	24.5 t
9	Rocket engine for maneuver of descent	0.55 tons	0.5 t
10	Landing gear	4.95 tons	4.5 t
11	Cargo bin	2.60 tons	2.5 t
12	Oxygen, food, and water for 9 explorers for 400 days on Mars	18.70 tons	17.0 t
13	2 caterpillar tractors	6.60 tons	6.0 t
14	1 fuel trailer	1.10 tons	1.0 t
15	Fuel for caterpillars	6.60 tons	6.0 t
16	Pneumatic tent, incl. combined air and water recuperation system, radio equipment, food-heating facility, cots, airlock, etc.	7.70 tons	7.0 t
17	Heating equipment or fuel for pneumatic tent	3.30 tons	3.0 t
18	10 space suits (1 spare)	1.10 tons	1.0 t
19	Research gear	5.50 tons	5.0 t
20	Tools	1.10 tons	1.0 t
21	Special gear for erection of nose section	2.20 tons	2.0 t
22	Total weight reduction	89.00 tons	81.0 t
Maneuver of re-ascent to circum-Martin orbit			
23	Thrust	110.0 tons	100.0 t
24	Initial weight ("earth weight"), referred to 1 g (# 3 —22)	76.0 tons	69.0 t
25	Empty weight (without payload)	7.3 tons	6.6 t
26	Payload of ascent (9 explorers plus 5.5 tons of research specimens collected on Mars)	6.5 tons	5.9 t
27	Propellant supply for maneuver of re-ascent	61.2 tons	55.6 t
28	Propellant supply left for maneuver of adaptation in orbit	1.0 tons	0.9 t
29	Final weight after completion of maneuver of re-ascent (# 25 + 26 + 28)	14.8 tons	13.4 t
30	Take-off acceleration (relative)	1.07 g	
31	Burning time	147 sec	
32	Cut-off altitude	78 mi	125 km
33	Cut-off velocity (horizontal)	2.3 mi/sec	3.7 km/sec
Maneuver of adaptation in circum-Martian orbit			
34	Thrust (half-throttle)	55 tons	50 t
35	Initial weight (# 29)	14.8 tons	13.4 t
36	Final weight (tanks exhausted)	13.8 tons	12.5 t
37	Burning time	4.8 sec	
38	Velocity increment for adaptation	590 ft/sec	180 m/sec
Dimensions			
39	Over-all length, body (complete landing craft)	130 ft	40 m
40	Rear diameter (cargo bin)	18 ft	5.5 m
41	Wing area	24,500 sq ft	2280 m²
42	Wing span	450 ft	137 m
43	Length of rocket for return to circum-Martian orbit	65 ft	20 m
44	Diameter of rocket	13 ft	4 m

TABLE IV. MAIN PERFORMANCE DATA OF RELIEF SHIP

		U.S.	METRIC
Initial weight of relief ship prior to departure from 2-hr orbit			
1	Landing weight of normal earth-returnable 4th stage (see Table I, # 75) *	10.34 tons	9.4 t
2	Normal landing payload (# 76)	—2.64 tons	—2.4 t
3	Wing (# 78)	—1.54 tons	—1.4 t
4	Empennage (# 79)	—0.33 tons	—0.3 t
5	Landing gear (# 80)	—0.55 tons	—0.5 t
6	Remote-control equipment (# 85)	—0.22 tons	—0.2 t
7	4th-stage rocket engine (# 90)	—0.11 tons	—0.1 t
8	Propellant tanks for rocket engine (# 91)	—0.11 tons	—0.1 t
9	Weight of "stripped-down" hull of 4th stage (# 1 —2 through 8)	4.84 tons	4.4 t
10	2-man relief crew	+0.22 tons	+0.2 t
11	Complete 3rd stage, incl. rocket engine (see # 67)	+2.32 tons	+2.1 t
12	7 extra 3rd-stage propellant tanks (see # 64, and footnote to # 72)	+5.39 tons	+4.9 t
13	Propellant manifolds for 7 extra tanks	+0.22 tons	+0.2 t
14	Dry weight of relief ship prior to departure (# 9 + 10 through 13)	12.99 tons	11.8 t
15	Propellant load—8 propellant tanks with a capacity of 25.2 tons each	201.60 tons	183.2 t
16	Initial weight of relief ship prior to departure (# 14 + 15)	214.59 tons	195.0 t
Main data on relief operation			
17	Radius of Mars ship's return orbit	56,000 mi	90,000 km
18	Period of revolution of return orbit	74 hr 10 min 0 sec	
19	Circular velocity of return orbit	1.32 mi/sec	2.12 km/sec
20	Perigee velocity of minimum ellipse between return orbit and 2-hr orbit	5.97 mi/sec	9.60 km/sec
21	Circular velocity of 2-hr orbit	4.40 mi/sec	7.07 km/sec
22	Velocity increment, maneuver 1 (# 20 —21)	1.57 mi/sec	2.53 km/sec
23	Mass ratio † required for # 22	2.47	

* All references in this table are to Table I.

† Mass ratios are based on a specific impulse of 285 sec, which has also been assumed for the Mars ships.

	U.S.	METRIC			U.S.	METRIC
24 Weight of relief ship after maneuver 1 (# 16 divided by 23)	86.9 tons	79.0 t	33 Weight of relief ship after maneuver 3 (# 30 divided by 32)		36.1 tons	32.8 t
25 Apogee velocity of connecting ellipse	0.537 mi/sec	0.865 km/sec	34 Seven peripheral tanks detached (see # 12)		—5.39 tons	—4.9 t
26 Velocity increment for maneuver 2 (# 19 —25)	0.783 mi/sec	1.255 km/sec	35 Weight of relief ship prior to maneuver 4 (# 33 —34)		30.71 tons	27.9 t
27 Mass ratio required for # 26		1.57	36 Velocity decrement maneuver 4 (see # 22)		1.57 mi/sec	2.53 km/sec
28 Weight of relief ship after maneuver 2 (# 24 divided by 27)	55.4 tons	50.3 t	37 Mass ratio required for # 36		2.47	
29 Weight of 12 returned explorers	+1.32 tons	+1.2 t	38 Weight of relief ship after return into 2-hr orbit (# 35 divided by 37)		12.42 tons	11.3 t
30 Weight of relief ship prior to maneuver 3 (# 28 + 29)	56.72 tons	51.5 t	39 Dry weight of relief ship, incl. payload (# 14 —12 + 29)		8.92 tons	8.1 t
31 Velocity decrement for maneuver 3 (see # 26)	0.783 mi/sec	1.255 km/sec	40 Propellant reserve left after relief mission (# 38 —39)		3.5 tons	3.2 t
32 Mass ratio required		1.57				

TABLE V. ORBITAL SUPPLY OPERATION ("SPACE LIFT")

	TONS U.S.	METRIC			TONS U.S.	METRIC
Dry cargo flights (payload 11 U.S. or 10 metric tons)			19 Total dry cargo for landing craft including payload in bin (# 13 to 18)		136.36	123.8
Passenger ship			*TOTALS*			
1 Dry weight (Table II # 115)	43.50	39.50	20 Total dry cargo carried by supply flights (# 8 + 12 + 19).		194.94	177.0
2 Oxygen, food and water for crew, Mars-bound voyage (II # 20, ftn.)	+8.10	+7.40	21 Total dry cargo supply flights		18 flights	
3 Payload after return (II # 108)	+12.76	+11.60	22 Total propellants for passenger ship (II, # 4 + 27 + 48 + 68)		1812.70	1648.10
4 Crew members (12) taken to departure orbit in earth-returnable ship (same as # 29; II # 104)	—1.32	—1.2	23 Total propellants for "interplanetary booster" (II # 4 + 27)		1632.30	1484.10
5 20-day oxygen, food, and water reserve for contact with relief ship (II # 106; included in # 13)	—1.20	—1.1	24 Total propellants for landing craft (# 15 + 16 + 17)		74.20	67.50
6 Specimens collected on Mars (II # 107)	—5.50	—5.0	25 Two transfer flights of relief ship from 2-hr. orbit to high orbit, and back. Twice total propellants for relief ship (IV, # 15)		403.20	366.40
7 8 rocket engines (44 U.S. tons thrust each, II # 111, 114 (These engines can be secured from abandoned 3rd stages of supply rockets in orbit, hence not "payload")	—8.80	—8.0	26 Propellants carried by orbital supply flights, # 22 through 25		3922.40	3566.00
8 Total dry cargo (# 1 through 7)	47.54	43.2	27 Total propellant supply flights, with 11.2 U.S. tons cargo each		350 flights	
Cargo ship			*Personnel flights **			
9 Weight of "interplanetary booster" after arrival in circum-Martian orbit (II # 121)	8.24	7.5	28 Weekly flight for rotation of assembly crews over 198 days		28 flights	
10 6 rocket engines and 4 tanks detached prior to maneuver 2 (II # 21, 22)	+11.60	+10.5	29 Ferrying of expedition members from ground to departure orbit		1 flight	
11 8 rocket engines (# 7)	—8.80	—8.0	30 Ferrying hull of relief ship		1 flight	
12 Total dry cargo without landing craft (# 9 through 11)	11.04	10.0	31 Return of expedition from 2-hour orbit to ground base		1 flight	
Landing craft			32 Total personnel flights		31 flights	
13 Weight after arrival in circum-Martian orbit, loaded and fueled (II # 122)	209.76	190.50	33 Unmanned flight for bringing landing craft hull to orbit; wings, etc., are flown as cargo		1 flight	
14 Oxygen, food and water for cargo ship crew, Mars-bound (II # 20)	+8.10	+7.40	34 Supply flights, total (# 21 + 27 + 32 + 33)		400 flights	
15 Propellants for maneuver of descent (III # 2 —3)	—12.00	—11.00	35 Total propellant consumption for "Space Lift," 1226 U.S. tons (1112 t.) per flight		490,000	445,000
16 Propellants for re-ascent (III # 27)	—61.20	—55.60				
17 Propellants for adaptation (III # 28)	—1.00	—0.90				
18 Empty weight without payload of wingless hull (III # 25). This hull is flown unmanned to orbit of departure like a standard third stage; see also # 33	—7.30	—6.60				

* In winged top stages with capacity of 14 men, including crew of 2. Available payload capacity of 1 ton per flight reserved for parts, tools, etc., during assembly.

BIBLIOGRAPHY

I. BOOKS AND IMPORTANT PAPERS ON MARS PUBLISHED PRIOR TO 1800

Bianchini (Blanchinus), Francesco. *F. B. veronensis astronomicae ac geographicae observationes, selectae ex ejus autographies.* Verona, 1737

Bose, G. M. *De Marte Conglaciante.* Leipzig, 1738

Cassini, Jean Dominique (Giovanni Domenico). *Martis circa proprium axem revolubilis observationes Bononiae habitae.* Bononiae, 1666

———. *Dissertatio apologetica de maculis Jovis et Martis.* Bononiae, 1666

Ehrenberger, B. H. *De Marte.* Coburg, 1738

Fontana, Francesco. *Novae Coelestivm terrestrivmq(e) rervm observationes.* Neapoli, 1646

Herschel, William. "Astronomical observations on the rotation of the planets round their axes, made with a view to determine whether the earth's diurnal motion is perfectly equable," *Philosophical Transactions,* vol. LXXI, 1781

———. "On the remarkable appearances at the polar regions of the planet Mars, the inclination of its axis, the position of its poles, and its spheroidical figure, with a few hints relating to its real diameter and atmosphere," *Philosophical Transactions,* vol. LXXIV, 1784

Kepler, Johannes. *Astronomia nova/ ΑΙΤΙΟΛΟΓΗΤΟΣ/ seu/ Physica coelestis/ tradita commentariis/ de motibus stellae/ MARTIS,/ ex observationibus C. V./ Tychonis Brahe/* Prague, 1609. A German edition appeared in Munich in 1929 under the title *Johannes Kepler's Neue Astronomie,* with an introduction by Max Caspar, who was also the translator.

Laurenti, Joannis Francisco de. *Observationes Saturni et Martis Pisaurienses.* Pisa, 1672

Maraldi, Giacomo Filippo. "Observations des taches de Mars pour vérifier sa révolution autour de son axe," *Historie et Mémoires de l'Académie des Sciences,* 1706

———. "Nouvelles observations de Mars," *ibid.,* 1720

Serra, Salvatore. *Martis revolubilis observationes romanae ab afficitis erroribus vindicatae.* Roma, Ex castro Sancti Gregorii, 1666

II. BOOKS AND IMPORTANT PAPERS ON MARS PUBLISHED FROM 1800 TO 1892

Backhuyzen, H. G. van de Sande. *Untersuchungen über die Rotationszeit des Planeten Mars und über Aenderungen seiner Flecke.* Leyden, 1885

Beer, Wilhelm, and Mädler, J. H. von. *Fragments sur les corps célestes du système solaire.* Paris, 1840. Articles by the two authors, comprising the body of this book, appeared in German in *Astronomische Nachrichten* from 1831 to 1842.

Boeddicker, Otto. "Notes on the physical appearance of the planet Mars, Birr Castle Observatory," *Scientific Transactions of the Royal Dublin Society,* 1882

Burton, C. E. "Physical observations of Mars, 1879–80," *Scientific Transactions of the Royal Dublin Society,* 1880

———. "Notes on the Aspect of Mars in 1882," *ibid.,* 1882

Cruls, L. *Mémoire sur Mars. Taches de la planète et durée de sa rotation.* Rio de Janeiro, 1878

Flammarion, Camille. *La Planète Mars, et ses conditions d'habitabilité.* Paris, 1892. Flammarion called this volume of 608 pages "synthèse générale de toutes les observations," and it actually is the most complete collection of all reports on Mars published before 1892. Since a second volume followed in 1909 (see section III), the 1892 volume was later referred to as volume I.

Flaugergues, Honoré. "Les taches de la planète Mars," *Journal de Physique,* LXIX, 1809

Franzenau, Felix von. "Mars im November 1864," *Sitzungsberichte der k.u.k. Akademie der Wissenschaften in Wien,* vol. LIII, Vienna, 1865

Green, Nathaniel E. "Drawings of Mars, made during the opposition of 1877 at Madeira," *Royal Astronomical Society, Memoirs,* vol. XLIX, 1877–1879

Gruithuisen, Franz von Paula. "Einige physisch-astronomische Beobachtungen des Saturns, Mars, des Mondes, der Venus, etc.," *Astronomisches Jahrbuch für 1817,* Berlin

Hall, Asaph. *Observations and Orbits of the Satellites of Mars.* Washington, D. C., 1878

Hoffmann, Dr. F. "Die neuesten Entdeckungen auf dem Planeten Mars," Heft 400 of the *Sammlung gemeinverständlicher wissenschaftlicher Vorträge,* herausgegeben von Rudolf Virchow and Fr. von Holtzendorff. Berlin, 1882. Originally a lecture delivered in January 1880.

Huggins, William, and Miller, A. "On the spectrum of Mars," *Philosophical Transactions,* 1864. This first attempt to use the spectroscope on the Martian atmosphere was made during the opposition of 1862.

Kaiser, J. "Untersuchungen über den Planeten Mars bei dessen Oppositionen in den Jahren 1862 und 1864," *Annalen der Sternwarte in Leiden,* vol. III, 1872

Lockyer, Joseph Norman. "Measure of the planet Mars, made at the opposition of 1862," *Memoirs of the Royal Astronomical Society,* vol. XXXII, 1863

Lohse, Otto. *Beobachtungen und Untersuchungen über die physische Beschaffenheit des Jupiter und Beobachtungen des Planeten Mars.* Observatorium zu Potsdam, 1878

————. "Beobachtungen und Untersuchungen über die physische Beschaffenheit der Planeten Jupiter und Mars," *Publikationen des astronomischen Observatoriums zu Potsdam,* vol. IX, 1882

————. "Beobachtungen des Planeten Mars," *ibid.,* no. 28, 1891

Schroeter, Johann Hieronymus. *Areographische Beiträge zur genauern Kenntnis und Beurteilung des Planeten Mars . . . Nach dem Manuscripte auf der Leidener Sternwarte herausgegeben von H. G. van de Sande Bakhuyzen, Director der Sternwarte.* Leyden, 1881. With 16 copper plates and 230 drawings.

Secchi, Father Angelo. "Osservazioni di Marte, fatte durante l'opposizione del 1858," *Memorie dell' Osservatorio del Collegio Romano,* Roma, 1859

————. "Osservazioni del pianeta Marte," *ibid.,* Nuova Serie, vol. II, Roma, 1863

Schiaparelli, Giovanni Virginio. *Osservazioni astronomiche e fisiche sull' asse di rotazione e sulla topografia del pianeta Marte.* Reale Accademia dei Lincei, Roma, 1878. Memoria seconda, Roma, 1881. Memoria terza, Roma, 1886

Schmick, Johann Heinrich. *Der Planet Mars, eine zweite Erde, nach Schiaparelli.* Leipzig, 1879

South, Sir James. "On the extensive atmosphere of Mars," *Philosophical Transactions,* 1831 and 1833

Terby, Dr. François. "Étude comparative des observations faites sur l'aspect physique de la planète Mars depuis Fontana (1636) jusqu'à nos jours (1873)," *Mémoires de l'Académie Royale des Sciences de Belgique,* tome XXXIX, 1875

————. "Études sur la planète Mars," *Bulletin de l'Académie Royale des Sciences de Belgique,* tome I, 1878

————. "Aspect de la planète Mars pendant l'opposition de 1879," *ibid.,* 1880

————. "Mémoire à l'appui des remarquables observations de M. Schiaparelli sur la planète Mars," *ibid.,* 1880

————. *Ensemble des observations faites à Louvain en 1888.* Brussels, 1889

Trouvelot, E. L. *The Trouvelot Astronomical Drawings Manual.* New York, 1882

————. "Four drawings of Mars," *Annals of the Astronomical Observatory of Harvard College,* vol. VIII, Cambridge, 1876

————. "Four drawings of Mars," in *L'Astronomie,* September 1884. See Plate VII.

Wislicenus, Walter. *Beitrag zur Bestimmung der Rotationszeit des Planeten Mars.* Leipzig, 1886

III. Books and Important Papers on Mars Published after 1892

Antoniadi, Eugenios Michael. *La Planète Mars.* Paris, 1930. Quarto, 240 pp., with 150 illustrations and 10 plates, mostly based on drawings made at the Observatory of Meudon, France. A very detailed study, discussing each Martian feature and its description by earlier astronomers.

————. "Mars Report for 1896–97" in *Memoirs of the British Astronomical Association,* 6, III, pp. 81 ff.; for 1903, *ibid.,* 16, IV, pp. 60 ff.; for 1905, *ibid.,* 17, II, pp. 37 ff.; for 1907, *ibid.,* 17, III, pp. 90 ff.; for 1909, *ibid.,* 20, II, pp. 44 ff.; for 1911–12, *ibid.,* 20, IV, pp. 163 ff.; for 1915–16, *ibid.,* 21, IV, pp. 90 ff.

Arrhenius, Svante. "The Planet Mars," in *The Destinies of the Stars.* New York and London, 1918. A translation of the Swedish edition of 1915. Most of the material and reasoning had been published by Arrhenius in German in 1911 as *Das Schicksal der Planeten.*

BIBLIOGRAPHY

Baumann, Adrian. *Erklärung der Oberfläche des Planeten Mars.* Zürich, 1909. 174 pp. with illustrations.

Brenner, Leo. *Mars-Beobachtungen 1896–1897, auf der Manora Sternwarte in Lussinpiccolo* (Istria). Berlin, 1898

Campbell, W. W. "Projections on the Terminator of Mars," *Publications of the Astronomical Society of the Pacific,* vol. VI, 1894

————. "A Review of the Spectroscopic Observations of Mars," *The Astrophysical Journal,* 1896, vol. I, pp. 255 ff.

Cerulli, Vincenzo. *Marte nel 1896–97.* Pubblicazioni dell' Osservatorio privato di Collurania. Teramo, 1898. 126 pp. with 2 plates and map.

————. *Nuove osservazioni di Marte (1898–1899); Saggio di una interpretazione ottica della sensazioni areoscopiche.* Teramo (Collurania), 1900. 140 pp. with illustrations.

Coblentz, W. W. *Temperature Estimates of the Planet Mars.* Scientific Papers of the Bureau of Standards, No. 512. U. S. Department of Commerce. Washington, D.C., 1925. 27 pp.

Dross, Otto. *Mars, eine Welt im Kampf ums Dasein.* Vienna and Leipzig, 1901. 171 pp., illustrations.

Dunham, Theodore, Jr. "Spectroscopic Observations of the Planets at Mount Wilson," *The Atmospheres of the Earth and Planets,* rev. ed., edited by G. P. Kuiper. Chicago: University of Chicago Press, 1951.

Fernández Ascarza, B. Victoriano. *El planeta Marte,* Madrid, 1924. 96 pp. including diagrams, maps, and tables.

Flammarion, Camille. *La Planète Mars et ses conditions d'habitabilité, Tome II.* Paris, 1909. 604 pp., with 426 illustrations and 16 maps. A continuation of the great work published in 1892, covering all observations from 1890–1901. A third volume was announced in the concluding chapter of this volume, but unfortunately was never published.

Henseling, Robert. *Mars; Seine Rätsel und seine Geschichte.* Stuttgart, 1925. 80 pp., 54 illustrations. A popular, condensed, but highly readable survey of the knowledge of Mars up to the time the book was written.

Housden, C. E. *The Riddle of Mars, the Planet.* London, 1914. 69 pp., 3 color plates, diagrams. An engineering study of the number of pipes and the horsepower required to handle the water from the melting polar caps, which are assumed to be 5 feet thick.

Joly, J. "On the Origin of the Canals of Mars," *The Scientific Transactions of the Royal Dublin Society,* August 1897. This paper tries to show that the "canals" are more or less in the nature of ridges, caused by the gravitational attraction of asteroids passing close over the Martian surface in very nearly horizontal paths prior to impact.

Kuiper, Gerard P. "Planetary Atmospheres and Their Origin," *The Atmospheres of the Earth and Planets,* rev. ed., edited by G. P. Kuiper. Chicago: University of Chicago Press, 1951

Livländer, R. "On the Colour of Mars," *Publications of the Astronomical Observatory of the University of Tartu* [Dorpat], vol. 27, no. 6, 1933

Lowell, Percival. *Mars.* Boston, 1895. 228 pp., 23 plates, map.

————. *Mars and Its Canals.* New York, 1906. 394 pp., illustrations and plates.

————. *Mars as the Abode of Life.* New York, 1909. 288 pp., illustrations and plates.

Lowell Observatory. *Bulletins,* No. I, June 1903, to date

————. *Annals,* Vol. I, 1898, to date

Manson, Marsden. "The climate of Mars," *Publications of the Astronomical Society of the Pacific,* vol. VII, 1895

Martz, E. P. "Variation in atmospheric transparency of Mars in 1939," *Publications of the Astronomical Society of the Pacific,* vol. 66, 1954

Menzel, Donald H. "The Atmosphere of Mars," *The Astrophysical Journal,* vol. 63, 1926

Menzel, Donald H., Coblentz, W. W., and Lampland, C. O. "Planetary Temperatures Derived from Water-Cell Transmissions," *ibid.*

————. "Temperatures of Mars, 1926, as Derived from Water-Cell Transmissions," *Publications of the Astronomical Society of the Pacific,* vol. 39, 1927

Moreux, Abbé Théophile. *La Vie sur Mars.* Paris, 1924. This short (94-page) study comes to the conclusion: *Mars nous présente l'état intermédiaire entre la Terre et la Lune et les phénomènes auxquels nous assistons de loin ne sont probablement que les dernières manifestations d'une vie qui s'éteint.*

Morse, Edward S. *Mars and Its Mystery.* Boston, 1913. 192 pp., plates and illustrations

Newcomb, Simon. "The Optical and Physiological Principles Involved in the Interpretation of the So-Called Canals of Mars," *The Astrophysical*

Journal, vol. XXVI, no. 1, July 1907. Report on experiments similar to Maunder's but with adults participating.

Pickering, William H. *Mars.* Boston, 1921. 174 pp., diagrams and plates. Collection of papers on Mars prepared by Professor Pickering between 1890 and 1914.

———. "Reports on Mars," Nos. 1-43, *Popular Astronomy,* 1914–1930. Running commentary on observations. Most libraries have bound sets of separates.

Plassmann, J. *Ist Mars ein bewohnter Planet?* Frankfurt am Main, 1901. 32 pp. Professor Plassmann answered the question of the title ("Is Mars an inhabited planet?") by stating that the "rays" on the moon and the "canals" on Mars might be the same unexplained natural phenomenon, but that nothing had been proved either way.

Richardson, Robert S. *Exploring Mars.* New York, 1954. 262 pp., illustrations and 16 plates. Not quite half of the book is devoted to a survey of the other planets of our solar system for purposes of comparing them with both Mars and earth.

Saheki, Tsuneo. "Martian Phenomena Suggesting Volcanic Activity," *Sky and Telescope,* February 1955

See, T[homas] J[efferson] J[ackson]. "Preliminary Investigation of the Diameter of Mars," *Astronomische Nachrichten,* vol. 157, 1901. A summary of the principal measurements made from 1651 up to 1901.

Schiaparelli, Giovanni Virginio. *Osservazioni astronomiche e fisiche . . . del pianeta Marte.* Memoria quattra. Reale Accademia dei Lincei, Roma, 1896

———. "The planet Mars," *Science,* May 5, 1899.

Slipher, Earl C. "Atmospheric and Surface Phenomena on Mars," *Publications of the Astronomical Society of the Pacific,* vol. 39, 1927

———. "New Light on the Changing Face of Mars," *National Geographic Magazine,* vol. CVIII, no. 3. September 1955.

Stoney, Johnstone. "Of Atmospheres upon Planets and Satellites," *The Scientific Transactions of the Royal Dublin Society,* November 1897

Struve, Hermann. "Ueber die Lage der Marsachse und die Konstanten im Marssystem," *Sitzungsberichte der Königl. Preuss. Akademie der Wissenschaften,* XLVIII, pp. 1056 ff., Berlin, 1911. A thorough discussion of the position of the axis of Mars, the orbits of its moons, etc.

Trumpler, R. J. "Observations of Mars at the Opposition of 1924," *Lick Observatory Bulletin,* vol. 13, Berkeley, 1927

Vaucouleurs, Gérard de. *The Planet Mars.* London, 1950. 88 pp., diagrams and plates. English edition of Vaucouleur's *La Planète Mars,* translated by Patrick A. Moore. Very fine survey of problems and of work done.

———. *Physics of the Planet Mars.* London and New York, 1954. 366 pp., diagrams and plates. English edition, with emendations by the author, of the original French *Physique de la Planète Mars.* This is the latest of the comprehensive French works on Mars; the "canals" are not mentioned except for a statement that they will not be discussed.

———. *Le Problème Martien.* Paris: Elzévir, 1946. 63 pp., 4 plates, diagrams.

Wallace, Alfred Russel. *Is Mars Habitable?* London, 1907. 110 pp., frontispiece. Written as a criticism of Percival Lowell's *Mars and Its Canals;* reaches the conclusion that the canals are not artificial and Mars is not habitable.

Webb, Harold B. *Observations of the Planet Mars.* Privately printed, 1936

———. *Observations of Mars and Its Canals.* Privately printed, 1941. Contains very many drawings of Mars and Martian features.

Whitmell, C. T. *The Moons of Mars.* (After 1900). This work is repeatedly mentioned by Antoniadi but is not listed by Flammarion (it presumably appeared after the conclusion of vol. II of *La Planète Mars*), and has not been seen by the authors.

Wright, W. H. "Photographs of Mars Made with Light of Different Colors," *Lick Observatory Bulletin,* XII, no. 366, 1925

———. "Filter Photographs of Mars," *Publications of the Astronomical Society of the Pacific,* vol. 51, 1939

IV. BOOKS ON RELATED SUBJECTS, PRINCIPALLY LIFE ON OTHER WORLDS

Bonestell, Chesley, and Ley, Willy. *The Conquest of Space.* New York: The Viking Press, 1949. 160 pp., 48 plates. Also in Dutch, Finnish, French, German, Italian, Japanese, and Swedish editions.

Braun, Wernher von. *The Mars Project.* Urbana, Ill.: University of Illinois Press, 1953. 92 pp., with diagrams. English edition of *Das Marsprojekt,* published as a special issue of the journal *Weltraumfahrt,* Frankfurt am Main,

1952. The first detailed study of the engineering requirements of a voyage to Mars.

Braun, Wernher von, et al.; ed. Cornelius Ryan, *Across the Space Frontier*. New York: The Viking Press, 1952.

———. *Conquest of the Moon*. New York: The Viking Press, 1953

Two symposia, the first dealing with the construction and uses of a manned space station, the second with an expedition to the moon.

Cyr, Donald Lee. *Life on Mars*. El Centro, Calif., 1944. 50 pp. with illustrations.

Dekker, Dr. Hermann. *Planeten und Menschen*. Stuttgart, 1926. 96 pp., with illustrations.

Flammarion, Camille. *Les mondes imaginaires et les mondes réels*. First published Paris, 1865; 21st ed. (598 pp.), Paris, 1892

———. *Les terres du ciel*. Paris, 1876. 600 pp.

———. *Le Pluralité des mondes habités*. Paris, 1877. 474 pp., colored frontispiece, diagrams.

Gramatzki, H. Jan. *Der Mensch und die Planeten*. Berlin, 1922. 159 pp., with illustrations.

Heuer, Kenneth. *Men of Other Planets*. New York: The Viking Press, 1954. 160 pp., illustrated.

Huyghens, Christiaan. *Kosmotheoros, sive De Terris Coelestibus, earumque Ornatu, Conjecturae*. The Hague, 1698. First edition in quarto with copper plates. Translations into various languages appeared during the eighteenth century.

Jones, Harold Spencer. *Life on Other Worlds*. New York, 1940. 299 pp., plates.

———. "Is There Life on Other Worlds?" *Smithsonian Report for 1939*; separately as Publication No. 3556. Washington, D. C., 1940.

Linke, Felix. *Die Verwandtschaft der Welten und die Bewohnbarkeit der Himmelskörper*. Leipzig, 1925. 165 pp., illustrations.

Lundmark, Knut. *Das Leben auf anderen Sternen*. Berlin, 1930. German edition, by Robert Henseling, of a Swedish work. 197 pp., illustrations and tables.

Maunder, E. Walter. *Are the Planets Inhabited?* London and New York, 1913. 166 pp., no illustrations.

Mercier, A. *Conférence astronomique sur la planète Mars*. Orléans, 1902. 48 pp. Printed report on a symposium held on April 23, 1900, in the Hôtel de Ville at Orléans, devoted to the proposition of forming a society for establishing communications with Mars.

Meyer, M[ax] Wilhelm. *Bewohnte Welten*. Leipzig, 1909. 94 pp., with numerous illustrations.

Moreux, Abbé Théophile. *Les autres mondes sont-ils habités?* Paris, 1926; rev. ed., Paris, 1950. 141 pp., with plates.

Nelson, The Earl. *There IS Life on Mars*. London: T. Werner Laurie Ltd., 1955. 142 pp. with plates. A popular discussion of life on other worlds, specifically Mars.

Papp, Desiderius. *Was lebt auf den Sternen?* Vienna, 1931. 345 pp., illustrations and plates. This is probably the last book in which intelligent inhabitants are taken for granted and the "oases" are matter-of-factly referred to as "the metropolitan cities of the Martian inhabitants."

Proctor, Richard Anthony. *Other Worlds Than Ours*. New York, 1870. 342 pp., diagrams.

Roy, Sharat Kumar. "The Question of Living Bacteria in Stony Meteorites," *Geological Series of Field Museum of Natural History*, vol. VI, no. 14, Chicago, 1935.

Rudaux, Lucien. *Sur les autres mondes*. Paris: Larousse, 1937. Folio, 223 pp., 313 illustrations, 73 black and white plates, 20 color plates.

Serviss, Garrett P. *Other Worlds: Their Nature, Possibilities and Habitability in the Light of the Latest Discoveries*. New York, 1901. 282 pp., plates and diagrams.

Strughold, Hubertus. *The Green and Red Planet: A Physiological Study of the Possibility of Life on Mars*. Albuquerque, N. M.: University of New Mexico Press, 1953. 108 pp., illustrations and plates. An interesting study of physiological and biological principles involved and the range of adaptability of terrestrial organisms in view of Martian conditions.

———. "Ecological aspects of planetary atmospheres, with special reference to Mars," *Journal of Aviation Medicine*, 23:130, 1952.

———. "Physiological Considerations on the Possibility of Life under Extraterrestrial Conditions," *Space Medicine*, ed. by John P. Marbarger. Urbana, Ill.: University of Illinois Press, 1951.

Younghusband, Francis. *Life in the Stars*. New York, 1928. 222 pp., with plates.

Whipple, Fred L. *Earth, Moon and Planets*. Philadelphia: Blakiston, 1941. 293 pp., well illustrated.

Note: For a complete bibliography of books on space travel in all languages see *Rockets, Missiles, and Space Travel* by Willy Ley (New York: The Viking Press, 1952).